Phytochemicals

Nutrient–Gene Interactions

Phytochemicals

Nutrient–Gene Interactions

Edited by

Mark S. Meskin
Wayne R. Bidlack
R. Keith Randolph

CRC Press
Taylor & Francis Group
Boca Raton London New York

CRC Press is an imprint of the
Taylor & Francis Group, an **informa** business
A TAYLOR & FRANCIS BOOK

CRC Press
Taylor & Francis Group
6000 Broken Sound Parkway NW, Suite 300
Boca Raton, FL 33487-2742

First issued in paperback 2019

ISBN-13: 978-0-8493-4180-9 (hbk)
ISBN-13: 978-0-367-39111-9 (pbk)

Library of Congress Card Number 2005054916

Library of Congress Cataloging-in-Publication Data

Phytochemicals : nutrient-gene interactions / edited by Mark S. Meskin, Wayne R. Bidlack, R. Keith Randolph.
    p. cm.
    Papers presented at the Fifth International Phytochemical Conference, "Phytochemicals: nutrient-gene interactions", Oct. 18 and 19, 2004, at California State Polytechnic University, Pomona.
    Includes bibliographical references and index.
    ISBN 0-8493-4180-9
    1. Nutrition--Genetic aspects--Congresses. 2. Genetic regulation--Congresses. 3. Gene expression--Congresses. 4. Nutrient interaction--Congresses. 5. Phytochemicals--Congresses. 6. Physiological genomics--Congresses. I. Meskin, Mark S. II. Bidlack, Wayne R. III. Randolph, R. Keith. IV. International Phytochemical Conference (5th : 2004 : California State Polytechnic University, Pomona)

QP144.G45P49 2006
612.3--dc22
          2005054916

Visit the Taylor & Francis Web site at
http://www.taylorandfrancis.com

and the CRC Press Web site at
http://www.crcpress.com

Preface

This is the fifth volume in a series of books that have emerged from five international phytochemical conferences held over the past decade at California State Polytechnic University, Pomona. The invited lectures at these conferences have ranged over a wide variety of phytochemical-related topics including discussions of a broad range of potentially important individual plant-based chemicals. The driving force behind these phytochemical conferences has been to promote authoritative scientific research to support the widespread renewed interest in phytochemicals and health that emerged in the 1990s, to identify gaps in the phytochemical knowledge base, to explore methodologies for screening and testing phytochemicals for efficacy and safety, to encourage pharmacokinetic studies of phytochemicals, and to look at the mechanisms of action of phytochemicals. The previous four volumes reflect the direction of phytochemical research during the past ten years and highlight the key phytochemicals and phytochemical classes that have shown promise in the promotion of health and the prevention of disease. The four volumes include *Phytochemicals: A New Paradigm* (1998), *Phytochemicals as Bioactive Agents* (2000), *Phytochemicals in Nutrition and Health* (2002), and *Phytochemicals: Mechanisms of Action* (2004).

The emphasis and direction of phytochemical research has shifted in the years since this series of international phytochemical conferences was initiated in 1996. Interest in phytochemicals has grown exponentially during the past few years as evidenced by the creation of research centers, the proliferation of newsletters, journals and books, and the many conferences and symposia that are now held on an annual basis. In order to maintain the relevance and value of the biannual conferences held at California State Polytechnic University Pomona, the organizers took a more focused approach to designing and planning the most recent conference. The previous conferences were quite eclectic. There was an identified theme but there were typically a number of diverse topics covered in order to keep abreast of all the important trends in phytochemical research.

The conference organizing committee decided that a critical direction for phytochemical research is the understanding of interactions between the wide variety of plant-based chemicals and genes. It was decided that the 2004 conference would focus primarily on phytochemical–gene interactions and the potential implications of those interactions for phytochemcial research, health care and research and development in the food and pharmaceutical/supplement industries. The study of nutrient–gene interactions is not new but research in this area is exploding with the completion of the human genome project and the availability of new screening technologies (see Chapter 3). There is significant epidemiological evidence (see Chapters 9 and 11) demonstrating that diets rich in fruits, vegetables, and whole grains can decrease the risk of chronic degenerative diseases. However, each person will respond to dietary components in a unique and individual way depending on his or her own genetic constitution. Understanding the interactions between phytochemicals and genes will begin to help deliver on the promise of truly individualized therapies tailored to one's genetic makeup. Such individualized care opens the door to the development of a wide range of new phytochemical-based therapeutic products.

The Fifth International Phytochemical Conference, "Phytochemicals: Nutrient–Gene Interactions," took place on October 18 and 19, 2004. The papers discussed at the conference, in expanded and updated form, are presented in this book. This book will be of value to both those who are relatively unfamiliar with the fields of nutrigenomics and nutrigenetics as well as those who want a current overview and update of these fields of research. Interactions between dietary factors and genes are explored in most of the chapters. These diet–gene interactions are discussed in the context of inflammation (Chapters 1, 4, 6, 10 and 12), cardiovascular and coronary heart disease (Chapters 2, 5, 8, 9, 10, 11 and 12), obesity (Chapters 3, 6, 7 and 11), type II diabetes mellitus (Chapters 3 and 11) and cancer (Chapters 10, 11 and 12). This book will appeal to nutrition researchers and nutritionists, food scientists and food technologists, pharmacists, botanists, and other allied health professionals who are interested in the opportunities created by an understanding of phytochemical–gene interactions.

In Chapter 1, Kornman and Fogarty define and describe nutrigenomics, nutrigenetics and pharmacogenetics. Inflammation, a critical component of atherosclerotic heart disease, is used as one example of how nutrigenomics can be applied in a way that will both individualize and improve treatment of disease. These authors point out that in the past nutrients were "first appreciated for their value as a fuel source," then as co-factors, and now "it is known that certain nutrients selectively alter gene expression through transcription factor systems that regulate the activation of specific sets of genes." Four mechanisms are offered for the nutrient effects on gene expression. Knowledge of nutrigenomics presents both opportunities and challenges for the development of therapeutic products. Kornman and Fogarty address the technical and ethical issues facing the development of phytochemical and nutritional products.

Ordovas points out in Chapter 2 that the current approach to preventive medicine is centered on blanket dietary recommendations such as the Dietary Guidelines for Americans or what he calls a "one-size-fits-all" approach. While this approach has seen some success, it clearly fails to take into account the well known individual variation in response to dietary interventions. This chapter examines "how genetic variations identified in samples from large population-based studies are beginning to provide hints about gene–diet and gene–environment interactions." The studies described by Ordovas have found significant interactions between factors in the diet, genetic variants and different markers of cardiovascular disease. With this type of information in hand it is beginning to be possible to identify individuals who may respond more favorably to one recommendation rather than another. In order to take advantage of this new information, phytochemical product development will require a solid grounding in nutrigenetics and nutrigenomics.

Chapters 3 and 4 discuss experimental methodologies. In Chapter 3, Kaput points out that "identifying genes regulated by diet and involved in chronic diseases is challenging because of the complexity of food and the genetic heterogeneity of humans." In his chapter, Kaput describes an experimental strategy that "identifies genes regulated by diet, genotype, and genotype X diet interactions." Utilizing this strategy, Kaput and his colleagues were able to identify 29 murine genes regulated by diet, genotype, or genotype X diet that map to diabetic Quantitative Trait Loci (QTL). The identified genes are likely to be associated with the development of type II diabetes mellitus in obese yellow mice. Clearly such studies need to be followed up in humans. However, it is also clear that a time will come when genetic testing will identify individuals who are likely to respond to specific diets or dietary components. In that future, Kaput notes "food companies may develop new markets and novel foods are likely to evolve in tandem with the ability to identify genotypes."

Regulations in the United States do not require dietary supplements, including phytochemical supplements, to be evaluated for safety and efficacy prior to marketing. In Chapter 4 Lemay makes a strong case that controlled clinical trials in the dietary supplement field are a necessary part of the product development process and are ultimately beneficial to supplement manufacturers. He describes a methodological approach that progresses from *in vivo* to *ex vivo* testing, followed by clinical evaluation. To demonstrate the first two steps in the process, Lemay reports on the results of a study that found that a hops extract dietary supplement exerted *ex vivo* Cox-2 inhibition comparable to that of a known pain-reliever. This is an important finding because Cox-2 selectivity might suggest an improved safety profile. Clinical evaluation of the extract will still be required to prove its effectiveness compared to other anti-inflammatory agents. Kornman comments in Chapter 1 that the lack of regulation of dietary supplements combined with the substantial pseudo-science being promoted to the consumer today can "poison the well" for science-based commercial products. In light of these comments, it is refreshing to read the recommendations of Lemay.

Interest in resveratrol, the major compound of the stilbene phytoestrogens found in grape skins, has been high ever since the French Paradox was described and moderate red wine consumption was hypothesized to provide antioxidant protection to consumers of high fat diets. Chapters in two previous volumes (Chapter 4 in *Phytochemicals in Nutrition and Health,* 2002; and, Chapter 9 in *Phytochemicals: Mechanisms of Action,* 2004) discussed the role of resveratrol in the prevention of cardiovascular disease. In light of the focus of the current volume on diet–gene interactions, the organizers thought it was important to revisit resveratrol, and discuss it in the context of lipid peroxidation and gene expression. In Chapter 5, Kutuk, Telci and Basaga indicate that "oxidatively modified low-density lipoproteins and end products of lipid peroxidation have all been shown to affect cellular processes by modulation of signal transduction pathways hence effecting the nuclear transcription of genes." In their review they discuss the molecular mechanisms that underlie resveratrol activity "with special focus on its effect on signaling cascades mediated by oxidized lipids and their breakdown products."

As indicated in Chapters 6 and 7, the worldwide prevalence of obesity and related disorders is reaching epidemic proportions and is increasing unabated. The phytochemical conference organizing committee realized the importance and complexity of this issue and invited two researchers to address the role of genes in the pathogenesis of obesity. Kern discusses adipose tissue gene expression in the context of inflammation and obesity in Chapter 6. One of his most interesting observations involves human evolution and the location of adipose tissue depots in the body. "Modern humans still have the genome of a hunter-gatherer, trapped in a body designed for a struggle against famine, infectious diseases and the threats from the elements. These Paleolithic threats seldom emerge, and the new threat is the result of the hunter-gatherer genome faced with an overabundant food supply and no need for physical activity." Pérusse echoes this theme in Chapter 7 where he places the obesity epidemic in the context of "a changing environment characterized by a progressive reduction in physical activity and the abundance of highly palatable foods. These changes in our lifestyle occurred over a period of time that is too short to cause changes in the frequencies of genes associated with obesity, which suggests that genes interacting with diet and other components of the modern lifestyle are important in determining an individual's susceptibility to obesity." Pérusse reviews gene–environment interactions in obesity and the implications of these interactions for the prevention and treatment of obesity.

The theme of the human genome incapable of evolving at a rate that could keep up with the breathtaking changes in the environment over the past 10,000 years since the introduction of agriculture and animal husbandry continues in Chapter 8 with a discussion of saturated fat consumption in ancestral diets and Chapter 10 which discusses the radical shift in the ratio of omega-6 to omega-3 fatty acids from pre-agricultural diets to modern Western diets. Cordain presents very provocative data in Chapter 8 that suggest the normal dietary intake of saturated fatty acids in our ancestral diet fell in the range of 10 to 15% of total energy, higher than current dietary recommendations of less than 10% of total energy. Based on these data Cordain claims that there is no genetic or evolutionary foundation for recommending such low levels of saturated fat intake and that we have insufficient data on the effects of long term low saturated fat intake. Simopoulos presents data in Chapter 10 that suggest Western diets are deficient in omega-3 fatty acids and too high in omega-6 fatty acids. She proposes that this imbalance contributes to cardiovascular disease, cancer, inflammatory and autoimmune diseases. In her chapter, Simopoulos discusses the influence of omega-3 and omega-6 fatty acids on the expression of genes involved in lipogenesis, glycolysis, inflammation, early gene expression, and vascular cell adhesion molecules.

Chapter 9 provides some of the background evidence suggesting a role for phytochemicals in health promotion and disease prevention. In this chapter Hu reviews and summarizes epidemiological research on plant-based foods and dietary patterns and concludes that there is substantial evidence "that healthy plant-based diets—those with adequate omega-3 fatty acids, that are rich in unsaturated fats, whole grains, fruits and vegetables—can, and should play an important part in the prevention of cardiovascular disease and other chronic diseases."

Slavin gives a comprehensive overview of the research on whole grains and chronic disease in Chapter 11. The protective components of grains are found in the germ and bran and include dietary fiber, starch, fat, antioxidant nutrients, minerals, vitamins, lignans and phenolic compounds. Consumption of the compounds in whole grains has been linked to reductions in risk of coronary heart disease, cancer, diabetes, obesity and other chronic diseases. Slavin makes a case for pursuing the mechanisms of protective action for these components of whole grains which might include phytonutrient-gene interactions.

Vitamin E is a popular topic in books on phytochemicals but these books typically focus on the antioxidant properties of vitamin E. Zingg and Azzi discuss the molecular activities of vitamin E in the final chapter, Chapter 12. What is unique about this chapter is that it focuses on the non-antioxidant cellular properties of vitamin E. In this comprehensive discussion, Zingg and Azzi cover the natural and synthetic vitamin E analogues, their occurrence in foods, plasma and tissue concentrations, uptake and distribution and finally the molecular action of vitamin E including the modulation of enzymatic activity and the modulation of gene expression. There are strong indications that each natural and synthetic vitamin E analogue can have a specific biological effect, often not associated with its antioxidant activity. The non-antioxidant activities might explain some of the health-promoting effects of vitamin E.

It is our hope that this volume will stimulate further interest and research in nutrigenomics and nutigenetics. A new understanding of phytochemical–gene interactions offers great potential for illuminating how diets rich in fruits, vegetables, and whole grains can decrease the risk of chronic degenerative diseases. The idea of a more individualized approach to dietary advice based on genotype is no longer science fiction. Safe and effective phytochemical products, based on genomics, can be developed responsibly based on real science.

Mark S. Meskin

Acknowledgments

The editors and authors wish to thank the Nutrilite Health Institute, Access Business Group, for its support of the 2004 Fifth International Phytochemical Conference, *Phytochemicals: Nutrient–Gene Interactions*, held in partnership with the Department of Human Nutrition and Food Science, College of Agriculture at the California State Polytechnic University, Pomona, October 18 and 19, 2004. The research presented at that conference contributed to the publication of this volume.

The editors would like to thank Susan B. Lee for all her support and constant encouragement of this project. We would also like to thank the editorial staff members at Taylor & Francis for their patience and excellent work, especially Marsha Hecht and David Fausel.

Editors

Mark S. Meskin, Ph.D., R.D., is professor and director of the Didactic Program in Dietetics in the Department of Human Nutrition and Food Science, College of Agriculture, California State Polytechnic University, Pomona. Dr. Meskin has been at Cal Poly Pomona since 1996.

Dr. Meskin received his Bachelor of Arts degree in psychology from the University of California, Los Angeles (1976), his master of science degree in food and nutritional sciences from California State University, Northridge (1983), and his Ph.D. degree in pharmacology and nutrition from the University of Southern California, School of Medicine (1990). In addition, he was a postdoctoral fellow in cancer research at the Kenneth Norris Jr. Cancer Hospital and Research Institute, Los Angeles (1990–1992). He received his academic appointment at the University of Southern California, School of Medicine (1992) and served as assistant professor of cell and neurobiology and director of the nutrition education programs (1992–1996). While at the University of Southern California, School of Medicine, he created, developed, directed, and taught in the master's degree program in nutrition science. Dr. Meskin has also served as a faculty member of the Department of Family Environmental Sciences at California State University, Northridge and the Human Nutrition Program at the University of New Haven, Connecticut.

Dr. Meskin has been a registered dietitian since 1984 and is also a certified nutrition specialist (1995). He has been involved with both the local and national Institute of Food Technologists for over 25 years. He is a past chair of the Southern California IFT and remains involved in SCIFT. Dr. Meskin has been an active Food Science Communicator for the national IFT, was a member of the IFT/National Academy of Sciences Liaison Committee, and has served as a member of the IFT Expert Panel on Food Safety and Nutrition. He is also involved in several IFT divisions, including the Nutrition Division, Toxicology and Safety Evaluation Division, and the Nutriceuticals and Functional Foods Division.

Dr. Meskin served as a science advisor to the Food, Nutrition and Safety Committee of the North American branch of the International Life Sciences Institute for a three-year term (2000–2002). He has been a long-time member of the advisory board of the Marilyn Magaram Center for Food Science, Nutrition and Dietetics at California State University, Northridge. Dr. Meskin was involved with the Southern California Food Industry Conference for many years as an organizer, chair, moderator, and speaker. He has also been a member of the Medical Advisory Board of the Celiac Disease Foundation.

Dr. Meskin is regularly invited to speak to a wide variety of groups and has written for several newsletters. He has been a consultant for food companies, pharmaceutical companies, HMOs, and legal firms. He is a member of many professional and scientific societies including the American Dietetic Association, the American Society for Nutrition, the American College of Nutrition, the American Council on Science and Health, the Institute of Food Technologists, and the National Council for Reliable Health Information.

He has several major areas of research interest including: (1) hepatic drug metabolism and the effects of nutritional factors on drug metabolism and clearance; nutrient–drug interactions; (2) the role of bioactive non-nutrients (phytochemicals, herbs, botanicals, nutritional supplements) in disease prevention and health promotion; (3) fetal pharmacology and fetal nutrition, maternal nutrition, and pediatric nutrition; (4) nutrition education; (5) the development of educational programs for improving science literacy and combating health fraud.

Dr. Meskin has co-edited four books on phytochemical research including: *Phytochemicals: A New Paradigm* (1998), *Phytochemicals as Bioactive Agents* (2000), *Phytochemicals in Nutrition and Health* (2002), and *Phytochemicals: Mechanisms of Action* (2004).

Dr. Meskin is a member of numerous honor societies including Phi Beta Kappa, Pi Gamma Mu, Phi Kappa Phi, Omicron Nu, Omicron Delta Kappa, Phi Upsilon Omicron, Gamma Sigma Delta, and Sigma Xi. He was elected a fellow of the American College of Nutrition in 1993 and was certified as a charter fellow of the American Dietetic Association in 1995. He received the Teacher of the Year Award in the College of Agriculture in 1999, the Advisor of the Year Award in the College of Agriculture in 2002, and the Advisor of the Year Award from Gamma Sigma Delta in 2004.

Wayne R. Bidlack, Ph.D., is dean of the College of Agriculture, at California State Polytechnic University, Pomona, and is professor with return rights in both the Department of Animal and Veterinary Science and the Department of Human Nutrition and Food Science.

Dr. Bidlack received his bachelor of science degree in dairy science and technology from the Pennsylvania State University (1966), his master of science degree from Iowa State University (1968), and his Ph.D. in biochemistry from the University of California, Davis (1972). In addition, he was a postdoctoral fellow in pharmacology at USC School of Medicine (1972–1974). He received his academic appointment at the University of Southern California (1974), served as assistant dean of medical student affairs (1988–1991), and served as professor and interim chair of pharmacology and nutrition (1992). Dr. Bidlack has also served as chairman and professor of food science and human nutrition, and as director of the Center for Designing Foods to Improve Nutrition at Iowa State University in Ames, from 1992 to 1995.

Dr. Bidlack has been a professional member of the Institute of Food Technologists for more than 20 years. He has served as a member of the annual program committee and has served as a member of both the Expert Panel on Nutrition and Food Safety and the Scientific Lectureship Committee, and as a scientific lecturer. He served as program chairman and chairman of the IFT Toxicology and Safety Evaluation Division (1989–1990) and has served as a member of the executive committee for both the TaSE Division and the Nutrition Division. He has served as editor of the TaSE newsletter. For the Southern California Section of IFT, Dr. Bidlack has served as councilor, chairman of the scholarship committee, program chairman and chairman of the section (1988–1989). He has also served as regional communicator for IFT in Southern California. Dr. Bidlack was elected a fellow of IFT in June 1998 and elected as counselor representative to the IFT executive committee.

Dr. Bidlack is past president of the Food Safety Specialty Section of the Society of Toxicology and has served on the International Life Sciences Institute Committee on Nutrition and Food Safety, and as a scientific advisor for the subcommittee on "Iron and Health" and the subcommittee on "Apoptosis" related to Fumonisin toxicity. He has also served as a member of the board for the CBNS and has actively contributed to the creation of the national certification exam. In addition, he is serving as abstract editor for the *Journal of the American College of Nutrition*. He served as the senior editor of the book *Phytochemicals as Bioactive Agents*, published by Technomic in January 2000.

In 1990, Dr. Bidlack received the Meritorious Service Award from the California Dietetic Association and the Distinguished Achievement Award from the Southern California Institute of Food Technologists. He was awarded honorary membership in the Golden Key national honor society in 1995 and in Gamma Sigma Delta in 1998. He also received the Bautzer Faculty University Advancement Award for Cal Poly Pomona, 1998.

His research interests are varied but integrate the general areas of nutrition, biochemistry, pharmacology, and toxicology. Specifically, his ongoing research projects examine the hepatic mixed function oxidase enzymes, drug and toxicant metabolism and conjugation; isolation and characterization of retinoic acid UDP-glucuronosyl transferase; nutrient-nutrient interactions between vitamins and minerals; and nutrient-xenobiotic toxicities. He also maintains interest in development of value-added food products, evaluation of biologically active food components (both plant and animal), and use of commodities for non-food industrial uses.

Dr. Bidlack has authored more than 50 scientific publications. For these efforts he has been elected to several national scientific societies, including the American Institute of Nutrition (American Society of Nutritional Science), the American College of Nutrition (certified nutrition specialist), the Institute of Food Technologists, the American Society of Pharmacology and Experimental Therapeutics, and others.

Dr. Bidlack has served the food industry as a consultant in a number of situations: Carnation, PER trials for new infant formulations; Hunt Wesson, PER Trials for low-fat peanut spread; Westcotek, various reviews of projects and product development proposals; Sunkist, article on nutrition; Excelpro, expert witness on protein chemistry and modification; Frito-Lay, expert witness on dietary fat and salt in the diet related to snack foods and on phytochemicals and functional foods; Apple Institute, article on Alar and food safety; International Life Sciences Institute, expert on nutrition and toxicology; Avocado Board, nutrition consultant; California Egg Commission, food safety; and others.

R. Keith Randolph, Ph.D., is senior group leader for the Department of Nutrition Science and Services at the Nutrilite Health Institute. His responsibilities include guiding the design and conduct of clinical testing and the sponsorship of nutrition-oriented symposia, programs, and distinguished lectureships. Dr. Randolph came to Nutrilite Health Institute in January 2000 with 17 years of combined experience in biochemical research and development as a lipid biochemist at the State College of New York, Stony Brook, the Medical College of Pennsylvania in Philadelphia, and the Cleveland Clinic Research Foundation in Cleveland, Ohio. He has published original research in the areas of lipid biochemistry, metabolism, nutrient–gene interactions, cardiovascular disease, and skin physiology. He holds a bachelor of science degree in chemistry and biology from Wayland College, Texas, and a Ph.D. in experimental pathology from the Bowman Gray School of Medicine of Wake Forest University, North Carolina. Dr. Randolph is a fellow of the American College of Nutrition, and holds memberships in the American Society for Nutritional Sciences, American Chemical Society, American Society for Molecular Biology and Biochemistry, and the American Organization of Analytical Chemists. Aside from scientific interests, he is an accomplished watercolor artist and has exhibited his work in New York and California. His artistic interest is on botanicals rendered in pigments he extracts from the plants he paints.

Contributors

Angelo Azzi, M.D.
Institute of Biochemistry and Molecular
  Biology
University of Bern
Bern, Switzerland

Huveyda Basaga, Ph.D.
Biological Sciences and Bioengineering
  Program
Faculty of Engineering and Natural
  Sciences
Sabanci University
Istanbul, Turkey

Loren Cordain, Ph.D.
Department of Health and Exercise
  Science
Colorado State University
Fort Collins, Colorado

Colleen Fogarty, M.S., R.D., L.D.N.
Nutrigenomics
Interleukin Genetics Incorporated
Waltham, Massachusetts

Frank B. Hu, M.D., Ph.D.
Department of Nutrition
Harvard School of Public Health
Boston, Massachusetts

Jim Kaput, Ph.D.
University of California at Davis
Davis, California
NutraGenomics
Chicago, Illinois

Philip A. Kern, M.D.
Central Arkansas Veterans HealthCare
  System
Division of Endocrinology
Department of Medicine
University of Arkansas for Medical
  Sciences
Little Rock, Arkansas

Kenneth Kornman, D.D.S., Ph.D.
Interleukin Genetics Incorporated
Waltham, Massachusetts

Ozgur Kutuk, M.D., Ph.D.
Biological Sciences and Bioengineering
  Program
Faculty of Engineering and Natural
  Sciences
Sabanci University
Istanbul, Turkey

Marc Lemay, Ph.D.
Nutrition Science and Services
Nutrilite Division of Access Business
  Group
Buena Park, California

Jose M. Ordovas, Ph.D.
Jean Mayer USDA Human Nutrition
  Research Center on Aging
Tufts University
Boston, Massachusetts

Louis Pérusse, Ph.D.
Department of Preventive Medicine
Laval University
Ste-Foy, Québec, Canada

Artemis P Simopoulos, M.D.
The Center for Genetics, Nutrition and
  Health
Washington, DC

Joanne Slavin, Ph.D.
Department of Food Science and
  Nutrition
University of Minnesota
St. Paul, Minnesota

Dilek Telci, Ph.D.
Biological Sciences and Bioengineering
  Program
Faculty of Engineering and Natural
  Sciences
Sabanci University
Istanbul, Turkey

Jean-Marc Zingg, Ph.D.
Institute of Biochemistry and Molecular
  Biology
University of Bern
Bern, Switzerland

Table of Contents

CHAPTER 1

Nutrigenomics: Opportunities and Challenges

Kenneth Kornman and Colleen Fogarty

CONTENTS

PHARMACOGENETICS: THE MODEL

Non-small cell lung cancer (NSCLC) is the number one cancer killer in the United States.[1,2] This type of lung cancer comprises 85% of all lung cancer cases. Iressa©, or gefitinib, is a well-tolerated drug prescribed for the treatment of NSCLC as a third line therapy — after the failure of two other well-established chemotherapy treatments. Although many oncologic treatments are combination therapies, Iressa has not been shown to enhance the effectiveness of other well-established drug therapies.[3]

Studies have demonstrated an overall response rate of 10%, which means 10% of the treated population will achieve a response. It is now known that essentially all of the people who respond to Iressa have a genetic mutation in the epidermal growth factor receptor (EGFR) gene.[4] Others have shown that the EGFR mutation occurs most frequently in Japanese people and is relatively uncommon in Caucasians.[5] Although, the positive clinical response rate to Iressa is relatively low, that

1

rate may be increased dramatically if patients are first screened for EGFR mutations. This is an example of pharmacogenetics (i.e., the use of genetic information to guide the use of drugs to patients who are most likely to have a favorable response).

The proceeding scenario is a good example of the impact our knowledge of human genetics and its variations can have on improving the pharmacologic treatment of disease. The field of pharmacogenetics offers practitioners and patients the opportunity to use genetic information to guide drug use and selection. It offers researchers and commercial drug developers the opportunity to more efficiently identify safe and effective drugs through a better understanding of individual differences. Additionally, as knowledge in the field grows, it will offer the opportunity to bring new effective drugs to market that may have been previously considered failures due to excessively variable responses when used in broad populations.

NUTRIGENETICS

The field of nutrigenetics employs concepts that are similar to pharmacogenetics, however, it is potentially more complex as it takes us away from disease toward a more wellness-oriented model of research and practice. The field of nutrigenetics contemplates the effects of genetic variations on nutrient influences in the health and disease of an individual. It allows us to identify the nutritional factors that are most influential for an individual to optimize their potential for a positive health trajectory throughout their life. Several terms, such as those in Table 1.1, are routinely used in discussions of how nutrients interact with the genome and the proteome. The terms *nutrigenetics* and *nutrigenomics* are often used interchangeably and may eventually be blended into one definition and one term. However, traditionally, these terms have different meanings. While nutrigenetics focuses on single gene variations and the nutritional implications associated with these variations; nutrigenomics

Table 1.1   Nutritional Genetics vs. Nutritional Genomics

What Is Nutritional Genomics?

Genomics
- Which genes and proteins are activated under different conditions

Genetics
- Mechanisms for inheriting specific traits
- How traits differ due to inherited factors

Nutrigenomics
- Effects of bioactive dietary compounds on the expression of genes (genome), proteins (proteome), and metabolites (metabolome)

Nutrigenetics
- Effects of genetic variations on nutrient influences on health and disease of an individual
- The nutitional factors that are most influential for a specfic individual

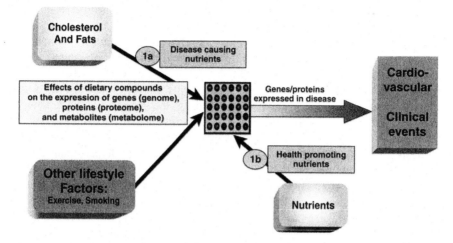

Figure 1.1 The paths of nutritional genomics.

contemplates the effects of nutritional compounds on the expression of multiple human genes, proteins, and metabolites.

Some of the most common applications of nutritional genomics are depicted in Figure 1.1. We know that dietary cholesterol and fats contribute to the clinical expression of cardiovascular disease. The use of "-omics" technologies, such as microarrays, to study which genes, proteins, and metabolites are activated by dietary fats in the development of atherosclerosis is one application of nutritional genomics (Figure 1.1;1a). One may also use the same technologies to attempt to find nutrients that counter-balance the effects of dietary fats on atherosclerosis (Figure 1.1;1b) as another application. Thus, nutritional genomics may be used to discover which molecular targets are influenced by specific nutrients to improve bioactive supplements or guide dietary modifications. This information may also be used, as in pharmacogenetics, to identify a subset of individuals who receive special benefit from specific nutrients, thereby reducing the variability of the response.

We should emphasize, however, the reality of the field in 2004–2005. Nutritional genomics does not mean that it is realistic to provide a unique formulation for each individual based on their genetics. It also does not mean that nutritional preparations or special diets can address all the component causes of a disease. Some applications of nutrigenomics and nutrigenetics do appear to be practical today. One is to substantially enhance our understanding of the genes and proteins that are activated by specific nutrients in specific situations. We already know of specific gene variations that influence single component causes of a disease, and we know that there are nutrient formulations or specific dietary considerations proven to benefit individuals who fit into certain genetic variation patterns associated with disease.

Inflammation Is One Example of Applied Nutrigenomics

When the body is presented with a challenge, such as a cut to the finger; the immune system responds by attacking foreign invaders and repairing the injured areas, sending immune cells into tissue in which they do not normally reside. The cut becomes *inflamed* as evidenced by the classic signs of redness and swelling due primarily to vascular dilation, leakage of fluid from the vessels, and emigration of leukocytes into the tissues. This state of acute inflammation is a normal and desired reaction of the body to this challenge. However, in some individuals, the inflammatory response to a challenge goes awry resulting in a chronic hyper-inflammatory state. Chronic inflammation can develop as a result of a persistent challenge or a dysregulated acute response. For example, if an individual has elevated low-density lipoprotein (LDL) cholesterol levels, it activates arterial endothelial cells to express adhesion molecules, such as VCAM-1 that recruit immuno-inflammatory cells. Monocytes in the circulation attach to the arterial wall, migrate into the wall, are activated, and phagocytose LDL-cholesterol. As they are activated, the monocytes release cytokines and growth factors that amplify and shape the inflammatory process in the arterial wall.

Inflammation and inflammatory mechanisms are now known to be central to many diseases, including cardiovascular disease, Alzheimer's disease, osteoporosis, obesity, and diabetes, which affect individuals in their mid-to-late years. Although inflammation-associated disease manifests itself later in life, inflammation can start much earlier and continue for many years without producing symptoms in the body. The inflammation association with common chronic disease has been most extensively developed for cardiovascular disease.

It is now known that atherosclerotic heart disease may develop from multiple paths, and that inflammation is a critical component. The first path in Figure 1.2

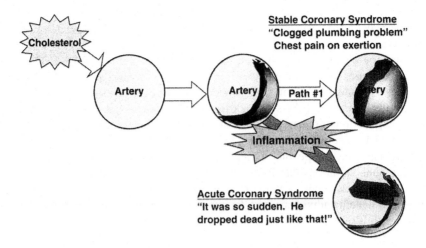

Figure 1.2   The paths of atherosclerotic heart disease.

demonstrates the interaction between cholesterol and the gradual increase in block-age, or stenosis, of the coronary arteries, which, if allowed to continue, will com-promise blood flow to the heart. An individual whose disease follows the trajectory of Path #1 will most typically experience chest pains during times of physical exertion. This path is an extension of the early atherosclerotic processes described above. After the monocytes enter the arterial wall, phagocytose lipid, and become activated macrophages, the growth factors and cytokines they release recruit other inflammatory cells to the area. This series of self-reinforcing processes produces an expanding accumulation of lipid-laden macrophages within the arterial wall. If the cholesterol challenge continues, the late stages of this process involves extension of the lipid deposits into the lumen of the artery. Thus, the atherosclerotic plaque begins to compromise blood flow, and this may lead to symptoms during physical exertion. Expansion of plaque into the artery lumen is detectable by angiography.

The alternate path is now known to lead to approximately half of first heart attacks in the U.S. That path involves different characteristics of inflammation in the walls of the coronary arteries that alter the physical and biological properties of the atherosclerotic plaques to make them less stable.[6,7] The unstable plaques have a less well-defined fibrous "cap," which is most likely due to a different balance of the cytokines and growth factors that determine the relative expression of inflam-mation vs. repair. Matrix-metalloproteinases are prominent in the unstable plaques, and they are therefore more prone to rupture. The acute results of the plaque rupture are that the lipid core is exposed to the blood elements, thereby inducing a thrombotic reaction, which can instantly block blood flow to segments of the heart. One of the challenges with the second path is that there is often no advance warning. In addition, an individual who follows this path may have cholesterol levels in the "normal" range, yet still be at risk for an acute cardiovascular event because of an increased propensity for inflammation. This exciting breakthrough in the understanding of cardiovascular disease has been supported by research findings on a blood marker of inflammation called *C-reactive protein* (CRP). The work of Paul Ridker has demonstrated that an elevated CRP level in people with a prior history of heart disease is as strong a predictor of future cardiovascular events as is elevated cho-lesterol. In addition, reduction of inflammation has been shown to have as beneficial an effect in reducing future events as lowering LDL cholesterol.[8]

The findings on CRP and cardiovascular disease risk have also revealed that some individuals maintain chronically higher levels of a low-grade inflammation than others. One of the important questions is what is different about the people with chronic inflammation, and can anything be done to modulate the inflammation and associated risk or disease. Cytokines are proteins or glycoproteins produced by most cells and are involved in the cell-to-cell communication involved in the immuno-inflammatory response, wound healing, and parts of energy metabolism. Interleukin-1 (IL-1) is a pro-inflammatory cytokine found at the beginning of the cellular communication cascade and, along with IL-6 and TNFα, plays a pivotal role in initiating and regulating inflammation.

IL-1 biologic activity is the result of primarily two agonists, IL-1α and IL-1β, and a naturally occurring antagonist, IL-1 receptor antagonist. Variations in these

genes are commonly found in many people in the population and have been associated with higher levels of inflammatory mediators, such as CRP.[9]

The presence of certain gene variations in the IL-1 gene cluster has also been
associated with the risk of heart disease. Individuals enrolled in the Atherosclerosis
Risk in Communities study who shared a common variation in the IL-1A gene
(homozygous for the rare allele at +4845) were found to be at a four-fold increased
risk of death due to an MI in the setting of low serum cholesterol.[10,11] In 1999,
Frances and colleagues demonstrated a strong association between the presence of
the rare allele in the IL-1RN gene variation at +2018 and IL-1B(-511) and coronary
artery disease.[12] The preceding studies provide evidence to support the evaluation
and modulation of individual genetic inflammatory tendencies early in life in order
to modulate these tendencies to prevent disease later in life. One route toward the
modulation of these tendencies lies in the path of nutrition.

Nutrients Directly Alter Gene Expression

Nutrients were first appreciated for their value as a fuel source for energy metabolism
and growth. In addition, there was extensive study of the roles of specific nutrients,
such as calcium, as essential elements of the structural components of the body. As
biochemical and physiological methods advanced, a new understanding emerged of
the role of certain nutrients as co-factors (i.e., vitamins) for the proper function of
enzymes involved in various aspects of metabolism and tissue maintenance. This
was a great advance in our understanding of the role of nutrition in biology and
represented a very different concept from the concept of "food as fuel." In recent
years, a third advance in the understanding of the role of nutrients has emerged. It
is now known that certain nutrients selectively alter gene expression through transcription factor systems that regulate the activation of specific sets of genes. The
remarkable aspect of this new role for certain nutrients is that it reveals specific
molecular targets for nutrients to influence the behavior of different tissues and under
different environmental conditions. Some nutrients bind to, or in some way directly
activate, specific transcription factors, which then regulate the activation of specific
sets of genes.[13] Polyunsaturated fatty acids are one example of nutrients that directly
alter transcription factors, such as the PPAR system. Other nutrients alter the oxidation-reduction status of the cell to indirectly influence transcription factor activity.
Many antioxidants will alter the activation status of the transcription factor NF-
kappa-β, which is a key regulator of many genes, such as those involved in several
aspects of the inflammatory response.

Common nutritional compounds, such as omega-3 fatty acids and isoflavones
have been shown to alter genes that code for cytokines, growth factors, cholesterol
metabolizing enzymes, and lipoproteins.[14–16]

Some of the strongest evidence for nutrient effects on specific gene expression
comes from animal studies of nutrient modification during the prenatal or early
postnatal periods. For example, investigators have shown in mice that addition of
specific nutrients to the normal diet of the mothers during the gestational period can
produce a life-long change in the color of the coat in the pups.[17–19] Others have
shown in rats that dietary modification of the pups only during the suckling period

can induce an adult onset obesity and hyper-insulinemia that is transmitted to the offspring of those pups.[20] These nutrient-gene effects appear to be "epigenetic" effects in which the nutrients alter the methylation and structure of the chemicals that influence how certain genes are available to be regulated; i.e., this is not a direct change of the DNA or an effect on regulatory elements within the DNA. These methylation effects may be induced by a short-term diet modification during a critical development phase, as in the study above, with diet modification during the suckling period.[20] As in this study, the genetic effect may not revert after changing the diet. This specific study involves an animal model with a genetic background that makes it susceptible to the epigenetic effects of the diet.

It should be emphasized that even though the effect of nutrients on gene expression may be prolonged and even extend to progeny, the nutrients do not change the DNA structure of genes. As far as we know, the nutrient effects on gene expression are through one of four major mechanisms. The first involves alterations in cell membrane chemistry such that eicosanoid metabolism is altered. Eicosanoids are key regulators of cell membrane receptor activity. Fatty acids on the *sn*-2 position of phospholipids in cell membranes are cleaved in response to specific stimuli via the action of phospholipase A2 to generate free fatty acids that are then substrates for eicosanoid metabolism primarily through the cyclooxygenase (COX), lipoxygenase, (LOX), and epoxygenase enzymes. The mixture of unsaturated fatty acids in the diet dramatically changes the distribution of different eicosinoid products. These eicosanoid products include prostanoids, such as prostaglandin E2, thromboxanes, leukotrienes, and hydroxylated eicosapentaenoate products. Diets enriched with omega-3 polyunsaturated fatty acids lead to a very different distribution of eicosanoid products compared to diets enriched with omega-6 fatty acids.[21] The different eicosanoid products lead to different gene expression patterns and different cell responses.

The second mechanism for nutrient effects on gene regulation involves direct interaction of nutrients with transcription factors that enter the nucleus and bind regulatory sites in the DNA to influence gene expression. Polyunsaturated fatty acids have been shown to influence expression of multiple genes by this mechanism.[22,23]

The third mechanism involves oxidation-reduction effects on transcription factors to alter their activation state. For example, the transcription factor NF-kappa-β that is involved in inflammatory mechanisms is activated by oxidation and is inhibited by dietary antioxidants.[24,25]

The fourth mechanism for nutrient effects on gene expression involves the epigenetic modification, such as methylation or acetylation, of the proteins that surround and stick to the DNA, thereby regulating how the gene functions. An example of dietary modification of epigenetic regulation of gene expression is given above.

Nutrient Effects May Be Determined by Genetic Differences in People

If nutrients affect expression of certain genes by directly or indirectly influencing transcription factors, one may easily postulate how a given nutrient may have very different effects in different individuals. For example, nutrients that influence transcription factors may have different effects in individuals with polymorphisms in the binding sites for the transcription factors that are affected by the nutrients. A recent study provided some support for this concept. One polymorphism in the inflammatory gene arachidonate 5-lipoxygenase (5-LOX) has been associated with risk for cardiovascular disease.[26] In general, individuals who are homozygous for the 5-LOX polymorphism have a greater thickness of the carotid arterial wall, one indicator of atherosclerosis burden. However, when the dietary intake of polyunsaturated fatty acids (PUFA) was taken into consideration, a strong interaction was observed between the PUFA intake and the 5-LOX polymorphism. Dietary intake levels of arachidonic acid, an omega-6 PUFA, had no effect on carotid wall thickness in subjects without the 5-LOX polymorphism, but increasing dietary levels of the omega-6 PUFA were significantly associated with increasing carotid wall thickness in subjects with 5-LOX gene variations. PUFAs are known to regulate expression of several inflammatory genes. It is reasonable to conclude from this study that the omega-6-PUFA activated the 5-LOX gene to a greater extent in individuals with the polymorphism than in those without the polymorphism, thereby increasing the risk for atherosclerosis in some individuals based on the presence of both dietary components and specific gene polymorphisms.

OPPORTUNITIES AND CHALLENGES WITH PRODUCT DEVELOPMENT

Companies and researchers who wish to introduce legitimate, science-based commercial products in the field of nutrigenomics face substantial challenges. Since these products are currently not well regulated, there is substantial pseudo-science being promoted to the consumer. Such products have little-to-no scientific basis but, unfortunately, have the potential to "poison the well" for others who are committed to the long-term opportunity to advance the science in this area. Unfortunately, current regulations do not provide the information that is necessary for even an intelligent informed consumer to differentiate legitimate products from the pretenders. Although physicians are very comfortable making decisions about the use of prescription drugs, they are generally not comfortable recommending nutritional products. This reticence comes primarily from the fact that the data supporting the value of specific nutritional products have been lacking, and it is difficult for a consumer to know whether a specific product will deliver the desired active ingredients in a bioavailable and safe formulation. With drugs, physicians normally rely on the Food and Drug Administration's certification of safety, efficacy, and bioavailable delivery of active ingredients. Such assurances do not currently exist for nutritional products, so physicians are naturally skeptical. The linkage of genetic testing to nutritional products introduces other appropriate concerns related to privacy, discrimination, and value of the test information.

If there is great potential to benefit the public with better nutritional products to enhance wellness, how can it be done responsibly? In our opinion, there are a few starting points. First, we must have adequately sized, randomized, controlled clinical trials to determine if the nutritional products are of real benefit. These studies must be "product-specific" (i.e., the consumer should be able to know that a specific product is capable of delivering the desired outcome) as opposed to "ingredient-specific" studies that do not really differentiate effective from ineffective products. The studies should be published in peer-reviewed journals. Second, we ultimately may need a third-party independent entity to establish standards for effective and safe nutritional products and certify that specific products meet the standards. Third, if genetic testing is part of the product, the issues associated with such testing must be addressed in a manner that provides some assurances to the consumer.

Effective nutritional products, based on genomics, will be developed based on real science and will be marketed responsibly by a few companies. Responsible companies and responsible researchers in this field should look for ways to identify the products that have true value, and find ways to assist the consumer in differentiating such products.

REFERENCES

1. American Lung Association, Trends in Lung Cancer Morbidity and Mortality, http://www.lungusa.org/atf/cf/%7B7A8D42C2-FCCA-4604-8ADE-7F5D5E762256%7D/lc1.pdf, 5, 2004.
2. American Cancer Society, Cancer Facts and Figures 2005, Atlanta, http://www.cancer.org/downloads/STT/CAFF2005f4PWSecured.pdf,13, 2005.
3. Veronese, M.L. et al., A phase II trial of gefitinib with 5-fluorouracil, leucovorin, and irinotecan in patients with colorectal cancer, *Br. J. Cancer*, 92, 1846, 2005.
4. Lynch, T.J. et al., Activating mutations in the epidermal growth factor receptor underlying responsiveness of non-small-cell lung cancer to gefitinib, *N. Engl. J. Med.*, 350, 2129, 2004.
5. Paez, J.G. et al., EGFR mutations in lung cancer: correlation with clinical response to gefitinib therapy, *Science*, 304, 1497, 2004.
6. Libby, P., Inflammation in atherosclerosis, *Nature*, 420, 868, 2002.
7. Libby, P., Atherosclerosis: the new view, *Sci. Am.*, 286, 46, 2002.
8. Ridker, P.M. et al., C-reactive protein levels and outcomes after statin therapy, *N. Engl. J. Med.*, 352, 20, 2005.
9. Berger, P. et al., C-reactive protein levels are influenced by common IL-1 gene variations, *Cytokine*, 17, 171, 2002.
10. Offenbacher, S. et al., Association of coronary heart disease with interleukin 1 gene variants in the Atherosclerosis Risk in Communities Study, in *The American College of Cardiology Annual Scientific Session*, Abstract No. 1028–166, 2004.
11. The ARIC Investigators, The Atherosclerosis Risk in Communities (ARIC) Study: design and objectives. The ARIC investigators, *Am. J. Epidemiol.*, 129, 687, 1989.
12. Francis, S.E. et al., Interleukin-1 receptor antagonist gene polymorphism and coronary artery disease, *Circulation*, 99, 861, 1999.
13. Muller, M. and Kersten S., Nutrigenomics: goals and strategies, *Nat. Rev. Genet.*, 4, 315, 2003.
14. Baumann, K.H. et al., Dietary omega-3, omega-6, and omega-9 unsaturated fatty acids and growth factor and cytokine gene expression in unstimulated and stimulated monocytes. A randomized volunteer study, *Arterioscler. Thromb. Vasc. Biol.*, 19, 59, 1999.

15. Fukushima, M. et al., Investigation of gene expressions related to cholesterol metabolism in rats fed diets enriched in n-6 or n-3 fatty acid with a cholesterol after long-term feeding using quantitative-competitive RT-PCR analysis, *J. Nutr. Sci. Vitaminol. (Tokyo)*, 47, 228, 2001.

16. Lamon-Fava, S., Genistein activates apolipoprotein A-I gene expression in the human hepatoma cell line Hep G2, *J. Nutr.*, 130, 2489, 2000.

17. Waterland, R.A., Do maternal methyl supplements in mice affect DNA methylation of offspring? *J. Nutr.*, 133, 238; author reply 239, 2003.

18. Waterland, R.A. and Jirtle R.L., Transposable elements: targets for early nutritional effects on epigenetic gene regulation, *Mol. Cell Biol.*, 23, 5293, 2003.

19. Waterland, R.A. and Jirtle R.L., Early nutrition, epigenetic changes at transposons and imprinted genes, and enhanced susceptibility to adult chronic diseases, *Nutrition*, 20, 63, 2004.

20. Srinivasan, M. et al., Programming of islet functions in the progeny of hyperinsulinemic/obese rats, *Diabetes*, 52, 984, 2003.

21. Mantzioris, E. et al., Biochemical effects of a diet containing foods enriched with n-3 fatty acids, *Am. J. Clin. Nutr.*, 72, 42, 2000.

22. Price, P.T., Nelson C.M., and Clarke S.D., Omega-3 polyunsaturated fatty acid regulation of gene expression, *Curr. Opin. Lipidol.*, 11, 3, 2000.

23. Tai, E.S. et al., Polyunsaturated fatty acids interact with the PPARA-L162V polymorphism to affect plasma triglyceride and apolipoprotein C-III concentrations in the Framingham Heart Study, *J. Nutr.*, 135, 397, 2005.

24. Shea, L.M. et al., Hyperoxia activates NF-kappaB and increases TNF-alpha and IFN-gamma gene expression in mouse pulmonary lymphocytes, *J. Immunol.*, 157, 3902, 1996.

25. Flohe, L. et al., Redox regulation of NF-kappa B activation, *Free Radic. Biol. Med.*, 22, 1115, 1997.

26. Dwyer, J.H. et al., Arachidonate 5-lipoxygenase promoter genotype, dietary arachidonic acid, and atherosclerosis, *N. Engl. J. Med.*, 350, 29, 2004.

CHAPTER 2

Gene–Diet Interactions, Blood Lipids, and Cardiovascular Disease Risk: The Rise of Nutrigenetics

Jose M. Ordovas

CONTENTS

INTRODUCTION

Current dietary recommendations for reducing cardiovascular disease (CVD) risk focus on mediating intermediate risk factors (e.g., hypertension, blood lipids, obesity, diabetes) believed to promote atherosclerosis. These recommendations are applied to the general population in a "one-size-fits-all approach" (with few exceptions for pregnancy and well-known diet–disease associations). To date, the effectiveness of these recommendations in lowering CVD prevalence has been limited. While it is difficult (and perhaps unwise) to pinpoint a single reason for this shortfall, it cannot be explained solely by a population-wide failure to follow the dietary guidelines. Experimental and observational studies have identified interindividual variability in response to various dietary modifications. The revised food pyramid provided by the United States Department of Agriculture (USDA) represents a small step toward addressing variability of dietary needs, but is still fails to account for the effects of age, gender, ethnicity, and genetic variations. The American Heart Association's (AHA) revised dietary guidelines in 2000 reflect an appreciation of the genetic and metabolic heterogeneity underlying the ability of nutritional guidelines to meet the needs of individuals.[1] AHA further emphasizes the need for understanding the influence of gene polymorphisms on individual responses to dietary factors.[2]

The current "one-size-fits-all" approach to preventive nutrition is centered around blanket dietary recommendations, such as reducing total cholesterol and triglyceride levels with a low-fat diet and regular, mild-to-moderate exercise. These recommendations are based on research indicating that serum total cholesterol concentrations are linearly associated with mortality from coronary heart disease (CHD),[3,4] but large between-country differences in CHD mortality rates indicate that other factors, such as diet, play a role in CHD risk. Previous guidelines recommending total cholesterol below 200 mg/dL to reduce CHD mortality were modified in light of a 26-year follow-up from the Framingham Heart Study in which 35% of CHD occurred in people with total cholesterol <200 mg/dL.[5] Further research determined the association between serum cholesterol and CHD mortality is primarily a function of increased low-density lipoprotein cholesterol (LDL-C) and that optimization of high-density lipoprotein cholesterol (HDL-C) is also important for reducing heart disease risk.

This progression in our understanding of the relationship between serum cholesterol and CHD mortality demonstrates the well-documented capacity for science to adjust its models based on new and more detailed information. In the field of nutritional genetics (nutrigenetics), researchers are studying gene–diet interactions in an effort to better understand factors mediating individual response to dietary interventions. The scientific literature is replete with accounts of interindividual variability in response to specific dietary factors (such as high or low intake of total fat or saturated fat). Exploration of the interactions between genetic variations and diet are beginning to reveal evidence that more individualized nutritional recommendations are required to address the interplay of dietary factors and genetic variations on risk of cardiovascular disease.

In this chapter, we explore how genetic variations identified in samples from large population-based studies are beginning to provide hints about gene–diet and gene–environment interactions that may lead to more individualized recommendations for preventing cardiovascular disease.

INTERACTION OF DIETARY FAT AND HEPATIC LIPASE GENE POLYMORPHISM ON HDL-C

The challenge of optimizing lipid profiles has frustrated lay people and researchers alike. To reduce the risk of developing heart disease, we're told to minimize LDL-C concentrations and maximize HDL-C concentrations. The inherent challenge, of course, is that while HDL-C can be increased by consuming saturated fatty acids (SFA), this strategy increases LDL-C even more, thus raising a person's risk of developing atherosclerosis.

Consider the case of two male neighbors who commute together to their jobs at the local utility company, each of whom has suffered a mild heart attack within the past year. The two have teamed up to make the necessary changes to mediate their atherosclerotic lipid profiles. Together they have quit smoking and are diligently consuming less saturated fat, exercising more regularly, and drinking red wine in moderation. But they are experiencing different effects of their modified lifestyles. Dave's doctor is very pleased with his progress. He has reduced his dietary fat intake substantially, which has resulted in lower LDL-C and increased HDL-C. His coworker, Gene, however, is frustrated. He's been doing the same things but his HDL-C remains low. The solution for Gene may be to start taking a lipid-lowering drug that increases HDL-C. Or is there a genetic component to his problem?

There exists a wealth of scientific research documenting this apparent contradiction. The heterogeneity of plasma lipid response to changes in dietary fat suggests a genetic component.[6-8] A retrospective analysis of data from 1,020 men and 1,110 women in the Framingham Offspring Study has revealed a genetic factor that may regulate plasma lipid response to dietary fat.[9] The lipolytic enzyme hepatic lipase (HL) plays an important role in HDL metabolism, such that overexpression of HL decreases HDL-C concentrations and deficiency of HL increases it.[10] At the genetic level, a common polymorphism in the hepatic lipase gene (LIPC) has been identified at position -514 bp upstream of the transcription initiation site for the hepatic lipase gene, in which a C-to-T substitution results in differing metabolism of HDL-C. For the T/T allele in the Framingham study, HDL-C increased with dietary fat intake <30% of total energy, but decreased when dietary fat was 30% of total energy. Among the C/C and C/T genotypes, however, no association was observed between the percentage of dietary fat and plasma HDL-C concentrations. Regression model analysis showed that in T/T individuals HDL-C decreased as total fat increased, but the opposite was true for C/C individuals. Thus, T/T homozygotes who reduce dietary fat intake to <30% of total energy will likely see improvements in their lipid profiles in the form of lower LDL-C and higher HDL-C. At first glance, this finding appears to conflict with the general observation that HDL-C increases with higher dietary fat intake. The apparent conflict is resolved, however, when one considers the high prevalence of the C/C and C/T genotypes in the general population. That is, for the majority of the general population (C/C and C/T), higher total fat intake does increase HDL-C, but the opposite is true for the minority T/T genotype. The Framingham study population described here, for example, consisted of 64% C/C, 33% C/T, and 3% T/T subjects.

This clear association between dietary fat intake and plasma HDL-C in carriers of the T/T allele provides strong support for a gene-nutrient interaction in the expression of HL on lipid metabolism. Based on these findings, T/T homozygotes would not be expected to gain an HDL-C benefit from dietary fat intake higher than 30% of total energy. If Dave is a T/T homozygote, then the low-fat diet is working well for him in terms of reducing LDL-C and increasing HDL-C. Like the majority of the population, though, Gene is not seeing an HDL-C benefit from his low-fat diet. Of course, we cannot discount the influences of other nutrients and gene polymorphisms on the overall lipid profiles of these two men.

This same study identified statistically significant interactions of saturated fatty acids (SFA) and monounsaturated fatty acids (MUFA) with LIPC genotype. However, no interaction was observed for polyunsaturated fatty acids (PUFA). When consumption of animal fat and vegetable fat were analyzed separately, a statistically significant interaction with LIPC genotype was present for animal fat but not vegetable fat.

The discovery that genetic profiles and dietary nutrients interact to determine lipid profiles is exciting, but there is still much to understand about how the multitude of dietary factors interact with specific genotypes to determine plasma lipid profile and other determinants of cardiovascular disease risk. Furthermore, the mechanism by which dietary fat interacts with the LIPC-514C/T polymorphism remains unknown. This study provides a plausible explanation for the interindividual variability in HDL-C response to dietary intervention observed anecdotally and in well-designed studies.

INTERACTIONS OF APOLIPOPROTEIN E (APOE) GENOTYPE WITH OBESITY AND ALCOHOL

Three major alleles of the APOE gene (APOE2, APOE3, and APOE4), which codes for the plasma protein apolipoprotein E (apoE), have been studied for their associations with plasma lipoprotein concentrations. Polymorphisms in the APOE gene have been shown to influence plasma lipoprotein concentrations in carriers of certain alleles. Specifically, the E4 allele has been associated with increased LDL-C, whereas the E2 allele has been linked with lower LDL-C.[11] Both E2 and E4 have been linked with increased plasma triglycerides.[12]

Again using data from the Framingham Offspring Study (1,014 men and 1,133 women), researchers examined the potential interaction of APOE and alcohol intake on LDL-C.[13] Carriers of the E2 allele (E2/E2 or E2/E3) typically have lower LDL-C than E3 homozygotes (E3/E3); E4 carriers (E3/E4 or E4/E4) tend to have higher LDL-C than E3 homozygotes. In this study, the expected lowering effect of the E2 allele on LDL-C was present in the overall study population. However, when stratified by alcohol intake (drinkers vs. nondrinkers), the expected effects of the E2 and E4 alleles were not observed in nondrinking men. The lowest LDL-C concentrations were observed in E2 male drinkers and the highest concentrations were seen in E4 male drinkers. In fact, male drinkers with the E2 allele had significantly lower LDL-C than male nondrinkers with the E2 allele. Conversely, among carriers of the E4 allele, LDL-C was higher in drinkers than in nondrinkers. There was no difference in LDL-C concentrations between drinkers and nondrinkers with the homozygous E3 genotype. In women, however, the expected effects of APOE genotype on LDL-C were observed in drinkers and nondrinkers alike.

Extensive research, including a meta-analysis of studies, suggests that moderate alcohol consumption (1 to 2 drinks per day) reduces CHD risk due to changes in lipid profile,[14–17] including increased HDL-C concentrations.[18,19] However, the meta-analysis did not examine LDL-C concentrations, and other research on the association of alcohol intake with LDL-C has been contradictory.[20–24] Evidence from this Framingham study of APOE genotypes suggests that the beneficial effects of alcohol intake on LDL-C may not apply to male carriers of the APOE4 allele.

Studies seeking a link between APOE4 genotype and insulin resistance or diabetes have failed to uncover a genotype–phenotype interaction.[25–27] However, another analysis of subjects from the Framingham Offspring Study tested the hypothesis that obesity modulates the association between APOE genetic variation and fasting insulin and glucose levels.[28] Researchers retrospectively analyzed genetic, biochemical, and body mass index (BMI) data for 2,929 men and women from the Framingham study. In this analysis, obese men with the E4 allele had higher plasma glucose and insulin levels than subjects with the homozygous E3 genotype. Furthermore, obesity was associated with higher fasting insulin in all three genotypes but with higher fasting glucose only in the APOE4 genotype. No associations between genotype and plasma glucose or insulin concentrations were observed among nonobese men or among women. Here is evidence of both a genotype (APOE4)-phenotype (obesity) interaction and a sex-specific association. In other research, interactions have been reported between obesity and APOE4 on plasma triglyceride concentrations.[12,29]

The "personalized nutrition" lesson from this study is that while weight control is important for mediating insulin sensitivity and reducing CVD risk, weight control may be particularly important for male carriers of the APOE4 allele to prevent increased fasting insulin and glucose (both risk factors for diabetes).

To put this in perspective, consider three male golfers heading into the clubhouse for a beer after a round of golf. The men represent the genotypes E2, E3, and E4, which correlate to approximately 14%, 65%, and 21% of the aforementioned study population, respectively. According to the research described above, only E2 will get the benefit of lower LDL-C by regularly partaking of a post-game beer. All else the same, E3 will have the same LDL-C whether or not he makes a habit of drinking a beer after regular golfing outings, and E4's LDL-C concentrations are likely to be higher if he typically drinks alcohol, even if only in moderation. When it comes to diabetes risk, E4 will also need to be especially careful to maintain a healthy weight, because mixing obesity with his APOE genotype will increase his risk for elevated serum insulin and glucose concentrations.

INTERACTION OF POLYUNSATURATED FATTY ACIDS (PUFA) AND APOA1 G-A POLYMORPHISM

While it is known that dietary fatty acids and G-to-A substitution at −75 base pair (bp) of the APOA1 gene can each influence plasma lipoprotein concentrations, studies examining the effect of each factor separately have produced conflicting results. In particular, evidence regarding the effect of dietary PUFA on HDL-C concentrations is conflicting. Traditionally, increased dietary PUFA has been thought to lower plasma HDL-C concentrations. However, analysis of allelic variations in the APOA1 gene and dietary PUFA in a subset of the Framingham Offspring Study challenge this association.[30] In this analysis of 755 men and 822 women for whom complete genetic, biochemical, and dietary information was available, the association of higher dietary PUFA with lower HDL-C concentrations was observed for female G/G homozygotes but not for female carriers of the A allele (A/A or G/A). Because the homozygous G/G genotype accounts for approximately 70% of the general population, it is not surprising that studies examining only the association between PUFA and HDL-C, without considering APOA1 genotype, would fail to detect an HDL-lowering effect of PUFA. This Framingham analysis uncovered the differential response to dietary PUFA between female carriers of the A allele (A/A or G/A) and female G/G homozygotes.

This study is particularly interesting because it describes a sex-specific differential effect of dietary PUFA on HDL-C concentrations. Female G/G homozygotes had higher HDL-C concentrations than female carriers of the A allele when dietary PUFA was <4% of total energy, but the opposite was true when dietary PUFA was >8% of total energy. In male subjects, the interaction between PUFA and G-A polymorphism was only statistically significant when alcohol consumption and smoking were factored into the regression model, confirming the importance of analyzing men and women separately.

In this scenario, higher PUFA intake benefits female carriers of the A allele by increasing their HDL-C concentrations and thus reducing their risk of cardiovascular disease. In addition, no difference was observed between genotypes for moderate dietary PUFA (4 to 8% of total energy), which is consistent with previous studies that reported no effect of dietary PUFA related to allelic variation. The findings from this study may explain the conflicting evidence from previous studies that did not examine gene–diet interactions in a sex-specific manner.

INTERACTION OF PUFA AND PPARA POLYMORPHISM ON PLASMA TRIGLYCERIDES AND APOC-III

Another gene–PUFA interaction identified by analyzing data from the Framingham Offspring Study is the relationship between PUFA intake and PPARA-L162V polymorphism. In this study, PUFA intake (predominantly in the form of vegetable oil) was associated with lower plasma triglyceride and apoC-III concentrations in carriers of the 162V allele (162V/162V or 162L/162V).[31] Peroxisome proliferator-activated receptor alpha (PPARA) regulates multiple genes involved in lipid metabolism.[32,33] Substitution of leucine to valine at the PPARA 162 locus has been shown to influence PUFA activation, with effects varying depending on plasma PUFA concentration.[34,35] Previous population-based studies detected an association between 162V polymorphism and increased total and LDL-C, but not triglyceride concentrations, despite evidence that PPARA plays a prominent role in regulation of triglyceride-rich lipids.[36] This lack of association is likely a result of the low prevalence of the 162V allele in the general population (14% in this study).

While some people may know their plasma triglyceride concentrations, it is unlikely that the average person knows his or her plasma apoC-III concentrations. Imagine, though, two female triathletes who are master's degree candidates in nutrition. They live together, eat healthful diets, and cook with plenty of olive and canola oil, believing it to be good for their lipid profiles. As triathletes and nutrition professionals, they of course abstain from smoking and drink alcohol only in moderation. One of the two, however, is concerned because her high plasma triglyceride concentrations persist despite her diet and exercise regimen. The answer to her conundrum may lie in the results of the aforementioned study evaluating PPARA-L162V polymorphism and PUFA intake.

In this retrospective analysis of data on 1,003 men and 1,103 women in the Framingham Offspring Study, higher PUFA intake (predominantly in the form of vegetable oil) was associated with lower plasma triglyceride and apoC-III concentrations in carriers of the 162V allele.[31] This benefit of increased PUFA intake was not observed in 162L homozygotes. Perhaps even more interesting, when the data were stratified by PUFA consumption greater than and less than the population mean (6% of total energy), a different effect was seen. Among subjects who consumed <6% of total energy from PUFA, triglyceride and apoC-III concentrations were slightly higher among carriers of the 162V allele than 162L homozygotes. The interactions observed were consistent whether PUFA intake was analyzed as a discrete or continuous variable and when controlling for sex, age, BMI, smoking, alcohol, diabetes, blood-pressure medications, estrogen therapy, and energy intake. Based on these findings, higher dietary PUFA appears to be very important for lipid metabolism in carriers of the 162V allele, but less important for 162L homozygotes.

Once again, gene–diet interactions for PUFA and the PPARA-162V genotype may explain the anecdotal evidence from our female triathletes. While any number of gene–diet-environment interactions may be at play in this case, perhaps the triathlete whose high triglyceride concentrations persist despite her high PUFA consumption is an LL homozygote who does not benefit from increased dietary PUFA. The challenge now is to determine what gene, diet, and environmental factors can be employed to reduce her triglyceride concentrations, or perhaps her genetic make-up even protects her from the cardiovascular risk associated with elevated triglycerides.

INTERACTION OF N-6 AND N-3 FATTY ACIDS WITH ARACHIDONATE 5-LIPOXYGENASE GENE PROMOTER

Research has identified differential effects of the n-6 and n-3 families of polyunsaturated fatty acids on plasma lipid profiles, inflammatory response and cardiovascular risk.[37] As the primary substrate for 5-lipoxygenase, n-6 and its metabolic precursor, linoleic acid, enhance the production of leukotrienes, which could induce arterial atherogenesis.[38–41] In addition, variants of the 5-lipoxygenase gene promoter may be linked with atherosclerosis.[42,43]

In the Framingham study of dietary PUFA interaction with triglycerides and apoC-III (described in the preceding section), n-6 and n-3 interacted with the PPARA-162V genotype to influence plasma triglycerides and apoC-III concentrations.[31] In that study, higher dietary n-6 reduced plasma triglycerides and apoC-III concentrations in 162V carriers. While the same was true for n-3 consumption, the increase was only statistically significant for the interaction with apoC-III concentrations, but not triglycerides. The 162L homozygotes did not experience a reduction in plasma triglycerides in response to increased n-6, but their triglyceride concentrations did decrease in response to increased n-3. This hypotriglyceridemic effect of n-3 has been reported consistently in experimental and observational studies, most likely due to the high prevalence of the homozygous L/L allele (86% in this study) at PPARA position 162 in the general population.

Retrospectively analyzing data on 470 men and women in the Los Angeles Atherosclerosis Study, researchers tested the hypothesis that carotid-artery intima-media thickness (a measure of atherosclerosis) is associated with arachidonate 5-lipoxygenase gene promoter (ALOX5) genotype and dietary intake of n-6 and n-3.[44] In this study population (55% non-Hispanic white, 30% Hispanic, 8% Asian-Pacific Islander, 5% black, and 2% other groups), carotid intima-media thickness was increased in carriers of two variant alleles of ALOX5 compared with carriers of the common allele. The magnitude of elevation was similar to that reported for diabetics in this study, suggesting that these variant alleles of ALOX5 carry an equivalent atherogenic risk. Higher dietary n-6 (in the form of arachidonic acid and linoleic acid) was associated with an additional increase in intima-media thickness in carriers of the variant alleles, thus compounding their already-elevated risk of atherosclerosis. Conversely, increasing dietary intake of n-3 reduced intima-media thickness in carriers of the variant alleles, in effect attenuating the atherosclerotic risk associated with the variant ALOX5 genotype. Intake of saturated fatty acids (SFA) and monounsaturated fatty acids (MUFA) had no effect on intima-media thickness for any of the ALOX5 alleles studied. Furthermore, dietary fatty acids of any kind (n-6, n-3, SFA, or MUFA) did not influence atherosclerotic profiles in carriers of the common allele.

CONCLUSIONS

The evidence for gene–diet interaction as an integral factor in modulating cardiovascular risk factors is mounting. There currently exists enough evidence to demonstrate real interactions among gene polymorphisms, diet, and health outcomes (disease). The research presented in this chapter begins to explain some of the interindividual variability observed anecdotally and documented in observational and experimental research. However, we are still a long way from implementing "personalized" dietary recommendations for various populations based on genetic variations.

It is important to stress the preliminary nature of these findings. In particular, five of the six studies were retrospective analyses conducted using the database of a large, population-based study of subjects with European white ethnicity. Furthermore, dietary intake data in these studies were based on subject-reported data in the form of a semi-quantitative food frequency questionnaire. Prospective intervention studies are needed to reproduce these observational findings and further explain the interactions among multiple gene polymorphisms, dietary factors, and other environmental factors (e.g., smoking, alcohol, exercise), as well as sex-specific differences.

As the nutrition industry enters this once predominantly academic arena of nutrigenetics, nutritional product development and dietary recommendations need to be based on solid science and draw on additional research from the sister field of nutritional genomics (nutrigenomics).

In closing, the research presented in this chapter emphasizes the importance of analyzing dietary factors and health outcomes according to genotype. When we conduct "one-size-fits-all" research, we tend to uncover only associations present in the predominant genotype and remain confused by contradictory findings. Fully characterizing the complexity of gene–diet interactions will require that we employ an individualized, or "one-size-does-NOT-fit-all," approach.

ACKNOWLEDGMENTS

I would like to express my most sincere gratitude to Laurie LaRusso, whose editorial contribution was essential for the completion of this chapter. Work presented in this chapter was supported by the National Heart, Lung and Blood Institute Contract N01-HC-25195, National Institutes of Health/NHLBI grant no. HL54776, contracts 53-K06-5-10 and 58-1950-9-001 from the U.S. Department of Agriculture Research Service.

REFERENCES

1. Krauss RM, Eckel RH, Howard B, et al., AHA dietary guidelines: revision 2000: a statement for healthcare professionals from the Nutrition Committee of the American Heart Association, *Circulation,* Oct 31 2000; 102(18):2284–2299.
2. Deckelbaum RJ, Fisher EA, Winston M, et al., Summary of a scientific conference on preventive nutrition: pediatrics to geriatrics, *Circulation,* Jul 27 1999; 100(4):450–456.
3. Verschuren WM, Jacobs DR, Bloemberg BP, et al., Serum total cholesterol and long-term coronary heart disease mortality in different cultures. Twenty-five-year follow-up of the seven countries study, *JAMA,* Jul 12 1995; 274(2):131–136.
4. Martin MJ, Hulley SB, Browner WS, Kuller LH, and Wentworth D., Serum cholesterol, blood pressure, and mortality: implications from a cohort of 361,662 men, *Lancet,* Oct 25 1986; 2(8513):933–936.
5. Castelli WP, Lipids, risk factors and ischaemic heart disease, *Atherosclerosis,* Jul 1996; 124 Suppl:S1-9.
6. Berglund L, Oliver EH, Fontanez N, et al, HDL-subpopulation patterns in response to reductions in dietary total and saturated fat intakes in healthy subjects, *Am J Clin Nutr,* Dec 1999; 70(6):992–1000.
7. Dreon DM, Fernstrom HA, Campos H, Blanche P, Williams PT, and Krauss RM, Change in dietary saturated fat intake is correlated with change in mass of large low-density-lipoprotein particles in men, *Am J Clin Nutr,* May 1998; 67(5):828–836.
8. Katan MB, Grundy SM, and Willett WC, Should a low-fat, high-carbohydrate diet be recommended for everyone? Beyond low-fat diets, *N Engl J Med,* Aug 21 1997; 337(8):563–566; discussion 566–567.
9. Ordovas JM, Corella D, Demissie S, et al., Dietary fat intake determines the effect of a common polymorphism in the hepatic lipase gene promoter on high-density lipoprotein metabolism: evidence of a strong dose effect in this gene-nutrient interaction in the Framingham Study, *Circulation,* Oct 29 2002; 106(18):2315–2321.

10. Santamarina-Fojo S, Haudenschild C, and Amar M, The role of hepatic lipase in lipoprotein metabolism and atherosclerosis, *Curr Opin Lipidol,* Jun 1998; 9(3):211–219.
11. Wilson PW, Myers RH, Larson MG, Ordovas JM, Wolf PA, and Schaefer EJ, Apolipoprotein E alleles, dyslipidemia, and coronary heart disease, The Framingham Offspring Study, *JAMA,* Dec 7 1994; 272(21):1666–1671.
12. Dallongeville J, Lussier-Cacan S, and Davignon J,, Modulation of plasma triglyceride levels by apoE phenotype: a meta-analysis, *J Lipid Res,* Apr 1992; 33(4):447–454.
13. Corella D, Tucker K, Lahoz C, et al., Alcohol drinking determines the effect of the APOE locus on LDL-cholesterol concentrations in men: the Framingham Offspring Study, *Am J Clin Nutr,* Apr 2001; 73(4):736–745.
14. Renaud SC, Gueguen R, Schenker J, and d'Houtaud A, Alcohol and mortality in middle-aged men from eastern France, *Epidemiology,* Mar 1998; 9(2):184–188.
15. Keil U, Chambless LE, Doring A, Filipiak B, and Stieber J, The relation of alcohol intake to coronary heart disease and all-cause mortality in a beer-drinking population, *Epidemiology,* Mar 1997; 8(2):150–156.
16. Rimm EB, Klatsky A, Grobbee D, and Stampfer MJ, Review of moderate alcohol consumption and reduced risk of coronary heart disease: is the effect due to beer, wine, or spirits, *BMJ,* Mar 23 1996; 312(7033):731–736.
17. Rimm EB, Williams P, Fosher K, Criqui M, and Stampfer MJ, Moderate alcohol intake and lower risk of coronary heart disease: meta-analysis of effects on lipids and haemostatic factors, *BMJ,* Dec 11 1999; 319(7224):1523–1528.
18. Glueck CJ, Hogg E, Allen C, and Gartside PS, Effects of alcohol ingestion on lipids and lipoproteins in normal men: isocaloric metabolic studies, *Am J Clin Nutr,* Nov 1980; 33(11):2287–2293.
19. Gaziano JM, Buring JE, Breslow JL, et al., Moderate alcohol intake, increased levels of high-density lipoprotein and its subfractions, and decreased risk of myocardial infarction, *N Engl J Med.* Dec 16 1993; 329(25):1829–1834.
20. Vasisht S, Pant MC, and Srivastava LM, Effect of alcohol on serum lipids & lipoproteins in male drinkers, *Indian J Med Res,* Dec 1992; 96:333–337.
21. Langer RD, Criqui MH, and Reed DM, Lipoproteins and blood pressure as biological pathways for effect of moderate alcohol consumption on coronary heart disease, *Circulation,* Mar 1992; 85(3):910–915.
22. Nakanishi N, Nakamura K, Ichikawa S, Suzuki K, and Tatara K, Relationship between lifestyle and serum lipid and lipoprotein levels in middle-aged Japanese men, *Eur J Epidemiol,* Apr 1999; 15(4):341–348.
23. McConnell MV, Vavouranakis I, Wu LL, Vaughan DE, and Ridker PM, Effects of a single, daily alcoholic beverage on lipid and hemostatic markers of cardiovascular risk, *Am J Cardiol,* Nov 1 1997; 80(9):1226–1228.
24. Rakic V, Puddey IB, Dimmitt SB, Burke V, and Beilin LJ, A controlled trial of the effects of pattern of alcohol intake on serum lipid levels in regular drinkers, *Atherosclerosis,* Apr 1998; 137(2):243–252.
25. Meigs JB, Ordovas JM, Cupples LA, et al., Apolipoprotein E isoform polymorphisms are not associated with insulin resistance: the Framingham Offspring Study, *Diabetes Care,* May 2000; 23(5):669–674.
26. Laakso M, Kesaniemi A, Kervinen K, Jauhiainen M, and Pyorala K, Relation of coronary heart disease and apolipoprotein E phenotype in patients with non-insulin dependent diabetes, *BMJ,* Nov 9 1991; 303(6811):1159–1162.

27. Shriver MD, Boerwinkle E, Hewett-Emmett D, and Hanis CL, Frequency and effects of apolipoprotein E polymorphism in Mexican-American NIDDM subjects, *Diabetes,* Mar 1991; 40(3):334–337.
28. Elosua R, Demissie S, Cupples LA, et al., Obesity modulates the association among APOE genotype, insulin, and glucose in men, *Obes Res,* Dec 2003; 11(12):1502–1508.
29. Reznik Y, Morello R, Pousse P, Mahoudeau J, and Fradin S, The effect of age, body mass index, and fasting triglyceride level on postprandial lipemia is dependent on apolipoprotein E polymorphism in subjects with non-insulin-dependent diabetes mellitus, *Metabolism,* Sep 2002; 51(9):1088–1092.
30. Ordovas JM, Corella D, Cupples LA, et al., Polyunsaturated fatty acids modulate the effects of the APOA1 G-A polymorphism on HDL-cholesterol concentrations in a sex-specific manner: the Framingham Study, *Am J Clin Nutr,* Jan 2002; 75(1):38–46.
31. Tai ES, Corella D, Demissie S, et al., Polyunsaturated fatty acids interact with the PPARA-L162V polymorphism to affect plasma triglyceride and apolipoprotein C-III concentrations in the Framingham Heart Study, *J Nutr,* Mar 2005; 135(3):397–403.
32. Sessler AM and Ntambi JM, Polyunsaturated fatty acid regulation of gene expression, *J Nutr,* Jun 1998; 128(6):923–926.
33. Fruchart JC, Duriez P, and Staels B, Peroxisome proliferator-activated receptor-alpha activators regulate genes governing lipoprotein metabolism, vascular inflammation and atherosclerosis, *Curr Opin Lipidol,* Jun 1999; 10(3):245–257.
34. Sapone A, Peters JM, Sakai S, et al., The human peroxisome proliferator-activated receptor alpha gene: identification and functional characterization of two natural allelic variants, *Pharmacogenetics,* Jun 2000; 10(4):321–333.
35. Flavell DM, Pineda Torra I, Jamshidi Y, et al., Variation in the PPARalpha gene is associated with altered function in vitro and plasma lipid concentrations in Type II diabetic subjects, *Diabetologia,* May 2000; 43(5):673–680.
36. Tai ES, Demissie S, Cupples LA, et al., Association between the PPARA L162V polymorphism and plasma lipid levels: the Framingham Offspring Study, *Arterioscler Thromb Vasc Biol,* May 1 2002; 22(5):805–810.
37. Wijendran V and Hayes KC, Dietary n-6 and n-3 fatty acid balance and cardiovascular health, *Annu Rev Nutr,* 2004; 24:597–615.
38. Spanbroek R, Grabner R, Lotzer K, et al., Expanding expression of the 5-lipoxygenase pathway within the arterial wall during human atherogenesis, *Proc Natl Acad Sci USA,* Feb 4 2003; 100(3):1238–1243.
39. Ferretti A, Nelson GJ, Schmidt PC, Kelley DS, Bartolini G, and Flanagan VP, Increased dietary arachidonic acid enhances the synthesis of vasoactive eicosanoids in humans, *Lipids,* Apr 1997; 32(4):435–439.
40. Kelley DS, Taylor PC, Nelson GJ, and Mackey BE, Arachidonic acid supplementation enhances synthesis of eicosanoids without suppressing immune functions in young healthy men, *Lipids,* Feb 1998; 33(2):125–130.
41. Lotzer K, Spanbroek R, Hildner M, et al., Differential leukotriene receptor expression and calcium responses in endothelial cells and macrophages indicate 5-lipoxygenase-dependent circuits of inflammation and atherogenesis, *Arterioscler Thromb Vasc Biol,* Aug 1 2003; 23(8):e32–36.
42. Mehrabian M, Allayee H, Wong J, et al., Identification of 5-lipoxygenase as a major gene contributing to atherosclerosis susceptibility in mice, *Circ Res,* Jul 26 2002; 91(2):120–126.
43. Aiello RJ, Bourassa PA, Lindsey S, Weng W, Freeman A, and Showell HJ, Leukotriene B4 receptor antagonism reduces monocytic foam cells in mice, *Arterioscler Thromb Vasc Biol,* Mar 1 2002; 22(3):443–449.

44. Dwyer JH, Allayee H, Dwyer KM, et al., Arachidonate 5-lipoxygenase promoter genotype, dietary arachidonic acid, and atherosclerosis, *N Engl J Med,* Jan 1 2004; 350(1):29–37.

GLOSSARY

Nutritional genetics	Field of nutritional research that examines the effect of genetic variations on the interaction between diet and disease
Nutritional genomics	Field of nutritional research that focuses on the effect of nutrients on the genome, proteome, and metabolome

Genetic Terms

Allele	Any of a series of two or more different genes that may occupy the same position or locus on a specific chromosome
ALOX5	Arachidonate 5-lipoxygenase gene promoter
APOA and APOE	Apolipoprotein A and Apolipoprotein E
Base pair	One of the pairs of chemical bases joined by hydrogen bonds that connect the complementary strands of a DNA molecule or of an RNA molecule that has two strands
Genotype	The genetic constitution of an individual referring to either a gene combination at one specified locus or any specified combination of loci
HL	Hepatic lipase
Homozygote	An individual with two identical genes at one or more paired loci in homologous chromosomes
LIPC	Hepatic lipase gene
Phenotype	Category or group to which an individual is assigned based on one or more clinical or observable characteristics that reflect genetic variation or gene–environment interaction
PPARA	Peroxisome proliferator-activated receptor alpha

Nutrition Terms

HDL-C	High-density lipoprotein cholesterol
LDL-C	Low-density lipoprotein cholesterol
MUFA	Monounsaturated fatty acids
PUFA	Polyunsaturated fatty acids
SFA	Saturated fatty acids

CHAPTER 3

Diet–Disease Interactions at the Molecular Level: An Experimental Paradigm

Jim Kaput

CONTENTS

INTRODUCTION

The Human Genome Project has provided the foundation for understanding health and disease at the molecular level. Genomes, however, respond to and interact with their environments. Nutrigenomics is an emerging field of research that examines the interactions between the nutritional environment and cellular/genetic processes. One of the primary aims of nutrigenomics is to understand the effects of diet on the activity of an individual's genes and health.[44] Nutritional genomics is an integrative systems biology that uses tools and concepts from nutritional science, molecular biology, genetics, and genomics. Since unbalanced nutrient intake alters the equilibrium between health and disease, nutrigenomics research assesses physiologies and pathologies. Much of the current emphasis of researchers is to associate physiological measurements (such as HDL, LDL, cholesterol, height, weight, enzyme activities, protein levels, metabolite concentrations, etc.) with genotype as measured by expression analyses or, more typically, single nucleotide polymorphisms (SNPs) in genes. The field is best summarized by the following tenets[42]:

- Improper diets in some individuals and under certain conditions are risk factors for chronic diseases.
- Common dietary chemicals alter gene expression and genome structure.
- The influence of diet on health depends upon an individual's genetic makeup.
- Some genes or their normal common variants are regulated by diet, and they may play a role in the development of chronic diseases.
- Dietary interventions based upon knowledge of nutritional requirements, nutritional status, and genotype can be use to develop individualized nutrition that optimizes health and prevents or mitigates chronic diseases. Optimal nutrition may also influence the aging process.

A BRIEF REVIEW OF NUTRITIONAL GENOMICS

Although Hippocrates proclaimed "let food be your medicine, and your medicine be your food," modern molecular biologists and geneticists usually do not include nutrients as a variable in studies of disease processes.[42] Even the simplest food contains hundreds of chemicals, some of which are nutritive (provide energy), non-nutritive but bioactive (e.g., a non-metabolized regulator molecule), or both nutritive and bioactive. Examples of bioactive chemicals are genistein and hyperforin. Genistein can be co-crystallized in the active site of the estrogen receptor — beta (ER-β),[52] a transcription factor that regulates a subset of estrogen-responsive genes. St. John's wort contains hyperforin, which has been shown to activate the pregnane X receptor (PXR).[99] PXR regulates members of the P450 gene family among other genes. Metabolism can also produce transcriptional ligands: certain lipids are converted to eiconosoids that bind to retinoid X receptors (RXRs) or peroxisome proliferator activated receptors (PPARs).[27,50] Several transcription factors are lipid sensor receptors regulating genes that metabolize lipid nutrients.[44]

Some dietary chemicals may also regulate signal transduction pathways. The polyphenol, 11-epigallocatechin-3-gallate (EGCG) inhibits tyrosine phosphorylation of the Her-2/neu receptor and epidermal growth factor receptor.[73] EGCG is found in green teas. Other dietary chemicals such as docosohexonoic acid (DHA), α-linolenic, linoleic, and oleic acids (among others) alter signal transduction pathways by changing the activity of G-coupled membrane proteins (e.g., GPR40) in a dose-dependent manner.[39]

Chemicals in food, therefore, can affect activities of proteins, receptors, and metabolic pathways, making it crucial to assess diet as a variable while conducting studies in model organism and humans. The need to test the effects of dietary chemicals applies not only to basic research, but also to diagnostic and drug development where diet might alter response to drugs or treatments.[42]

The Influence of Genetic Makeup on Dietary Responses

Dietary reference intakes (DRIs) are average intakes for the majority of individuals (~97%) in a population. However, these guides are often interpreted as optimum for a given individual under the false assumption that everyone is culturally, socioeconomically, physiologically, and genetically identical.[66] Each of these variables may alter DRIs independently or in combination, and assessing their impacts is one of the goals of nutrigenomics research. Much current research is focusing on the contribution of genetic variation to disease susceptibility and responses to diet.

Humans trace their genetic ancestry to Africa[47] and are 99.9% identical at the gene sequence level. Differences in phenotype such as hair and skin color, height and weight potential, and susceptibility to disease or health, are produced by the 0.1% variations in DNA sequence. These polymorphisms are responsible for changes in protein activity levels and gene expression levels.

- The canonical example of how single nucleotide polymorphisms (SNPs) can alter gene expression is a polymorphism in the promoter region of lactase-phlorizin hydrolase gene (*LCH* locus) that allows its expression into adulthood. Whereas mammals generally do not drink or metabolize lactose after weaning, mutations in the promoter of *LCH* gene in humans that occurred about 10,000 years ago in a northern European allowed expression beyond weaning into adulthood. The specific mutation most highly associated with lactase persistence is a C-13910T SNP located 14kb upstream of the *LCH* gene.[19] This polymorphism is thought to alter regulatory protein-DNA interactions controlling expression of the *LCH* gene.[37] Since milk is a rich source of nutrients, it may help prevent dehydration under draught conditions, and improved calcium availability, the mutation provided a selective advantage to carriers and was spread through the population. This C-13910T allele occurs at 86% frequency in the northern European population but at only 36% in southern European populations. Regulatory SNPs (rSNPs) in other promoters are likely to play a role in regulating gene expression.[2,6,53]

- Gene expression may also be affected post-transcriptionally by altering mRNA levels. The insulin receptor has at least two splice variants termed type A or B. Exon 11 in the insulin receptor gene is spliced from the mRNA in the Type A variant and is associated with hyperinsulinemia.[38,90] Over 30,000 alternative splice sites have been identified in a genome-wide analysis of humans,[56] although functional polymorphisms at these sites have yet to be fully characterized.

- Xenobiotics are often metabolized by the cytochrome P450 enzymes. The *CYP3A4* gene illustrates how SNPs in the coding sequence alters biochemical activities of enzymes, proteins, and cellular processes. Eighteen coding SNPs have been found in this gene[53] and six affect activity of the enzyme.[53] These include *CYP3A4*3* (M445T), *CYP3A4*7* (G56D), *CYP3A4*9* (V170I), *CYP3A4*10* (D174H), *CYP3A4*11* (T363M), and *CYP3A4*19* (P467S). These polymorphisms only slightly alter testosterone, progesterone, or 7-benyloxy-4-trifuloromethyl coumarin metabolism when compared to the activity of the reference sequence. Four other SNPs have more dramatic effects on enzyme activities: CYP3A4*2 (S222P) has a six-fold increase in K_M with decreased clearance for nifedipine; CYP3A4*12 (L373F) has increased efficiency of hydroxylations at 15β, 2β-positions of testosterone compared to the reference sample; CYP3A4*18 (L293P) had a modest increase of 6β hydroxylation of testosterone; CYP3A4*17 (F189S) had increased activity of chlorpyrofos de-sulfation.[53] Hence, xenobiotic, steroids, and drug metabolism will differ among individuals with variations in this gene.

Genetic diversity within a population adds statistical noise to studies of diet in humans. Variations in response to diet are caused by differences in gene expression patterns,[106] protein, and enzyme activities. Naturally occurring dietary chemicals may interact with the same classes of proteins in each individual, but because individuals have different variants of these proteins, the responses among individuals differ.

Since each individual is genetically unique and it is difficult to control the environment of free living organisms, humans are not good subjects for certain research projects. Although individual variability has been recognized for centuries, a 1956 book summarized variability in physiologies measured with modern bio-chemical tools.[102] More recently, high throughput gene expression technologies showed a high degree of inter- and intra-individual variation in peripheral blood samples.[75,100] Laboratory animals, on the other hand, provide a useful model for nutritional genomics studies because genotype and environment may be more rig-orously controlled. For example, inbred mice are genetically identical because of brother-sister matings over 20 generations.[60] Experiments can therefore be replicated. Rodents also contain 99% of all human genes,[71] and disease physiologies are gen-erally similar among mammals.

Certain Diet-Regulated Genes Are Involved in Disease Processes

Phenotypes result from the expression of genetic information. Qualitative and quan-titative changes in gene expression therefore contribute to the manifestation of the disease state. Epidemiological data from migration studies have demonstrated that changes in environment (diet) produce dramatic changes in disease incidence and severity. Hence, a subset of genes regulated by diet must be involved in disease initiation, progression, and severity.[45,68] The clearest example of genotype-diet inter-actions in chronic disease is type 2 diabetes, a condition that frequently occurs in sedentary, obese individuals, and certain minority groups.[4,5] Some individuals can control symptoms by increasing physical activity and by reducing caloric (and specific fat) intake.[64] Other chronic diseases do not show the phenotypic plasticity seen in some type 2 diabetics, that is, symptoms are not reversible. Although the molecular mechanisms are not yet established, chromatin remodeling and changes in DNA methylation induced by unbalanced diets are possible mechanisms that contribute to irreversible gene expression changes.

METHODS FOR STUDYING NUTRIGENOMICS

Molecular Approach

Understanding the molecular mechanisms whereby diet alters health requires analyses of diet-regulated gene expression. Goodridge and coworkers[30,63] were among the first to pioneer this approach, albeit, one gene at a time. Gene expression analyses are now a standard method for analyzing the effects of diet,[10,13,15,20,29] including caloric restriction.[8,57,58,74] Changes in gene expression are then associated with phenotype and can be explained by genetic variants in nuclear receptors, cis-acting elements in promoters, or differences in metabolism that produce altered concentrations of transcriptional ligands.

The limitations of assessing regulation of individual or multiple genes by diet are (i) determining cause from effect for each gene, that is, what is the subset of causative genes for a given phenotype, and (ii) gene expression patterns in one strain (or genotype) may be unique to that genotype. The results from inbred mouse strains[43] would suggest that individual humans[106] may have unique patterns of gene expression depending upon their genotype and diet. Individual qualitative and quantitative differences will complicate attempts to find patterns in gene expression results for dietary intake. With diet recalls being imprecise and controlling diets difficult in large population studies, identifying these complex interactions will be challenging.

The presence of a disease can be considered an additional environmental influence that could affect gene expression patterns. For example, obesity unmasks additional type 2 diabetes loci in C57BL/6 and BTBR mice,[89] which is caused by two interacting loci that affect fasting glucose and insulin levels — these effects were observed only in obese mice. Alleles from the two parental strains (C57BL/6 and BTBR) had different affects on the diabetic subphenotypes. Hence, one would predict changes in gene expression based upon the presence or absence of disease processes and changes caused by dietary differences. Separating these variables will be an important component of future experimental designs for determining the effect of diet on susceptibility and disease progression.

Genetic Approach

Genetic epidemiological studies have been successful in identifying gene regions or the causative gene when the disease is caused by a mutation in a single gene.[41] However, chronic diseases are the result of multiple genes, some of which interact with the environment. Gene–gene (epistasis) and gene–environment interactions complicate the search and identification for causative genes.[36] Genetic epidemiological associations of individual polymorphisms with chronic diseases or responses to diets are frequently not replicated in subsequent studies. Several explanations have been suggested, such as small sample size, poorly matched control groups, population stratification, and over-interpreting data.[9,54,79,93] These methods and approaches are being improved to eliminate such errors and to reliably identify genes involved in chronic diseases.[11,17,62,69,72,78]

In addition to the design issues, many genetic mapping techniques and gene association studies do not account for the effects of different environmental variables, such as diet, on the expression of genetic information.[42] Lander and coworkers[70] showed the importance of environment on expression of phenotypic traits in F2 and F3 generation tomato plants grown in Davis or Gilroy, CA, compared to plants grown near Rehovot, Israel. Only 4 of a total of 29 quantitative trait loci (QTL) were found at all three sites, and 10 QTL were found in only two sites. QTL identify multiple regions within all chromosomes that collectively contribute to a complex phenotype.[7,24,81] Since an individual QTL may encode many genes, identifying the causative genes within the QTL is challenging if genetic methods alone are employed.[93] Genes within any QTL contribute different amounts (~1 to 100%) to the disease process. The genes at each locus are likely to have alleles that contribute to the disease (deleterious alleles), neutral alleles, or beneficial alleles. The most severe case would result if deleterious alleles or variants are present at each locus so the total genetic contribution is 100%. The particular mix of deleterious, neutral, and beneficial alleles at different QTLs that a person inherits explains the differences in genetic susceptibility found in outbred populations such as humans. QTLs in different species often encode the same set of genes, suggesting that laboratory animals are good models for some human diseases. For example, eight loci in humans that are associated with diabetes or subtypes of diabetes[25] are syntenic to loci in mice associated with type 2 diabetes mellitus (T2DM) subpheontypes.

Genetic epidemiology studies of gene–environment interactions are compounded by the same factors that affect molecular (gene expression) analyses: epistatic interactions between genes (the obesity example in mice[89]) *in utero* effects, diet–gene interactions, and the "environmental history," that is, the life-long exposure to changing diets may alter expression of genetic information later in life.[88] Altered phenotypes in laboratory and farm animals may be produced by differences in maternal nutrition, which alter DNA methylation status,[12,14,40,101] and such epigenetic phenomena will likely affect gene-disease association studies. Sing and colleagues[88] discuss these multiple influences in an excellent review of changing expression of genetic information during aging.

IDENTIFYING DIET-REGULATED GENES INVOLVED IN DISEASE PROCESSES

Our laboratory developed an experimental strategy of comparing across genotypes and diets to identify genotype X diet interactions.[18,43,67,68,92] Our method is based on identifying differences in gene expression based on the differential susceptibility to diet-induced disease (Figure 3.1). For example, strain A is genetically susceptible to a diet-induced disease when fed an experimental but not control diet, and strain B would be refractory to both experimental and control diets. Animals are symptom-free when gene expression is analyzed because disease processes themselves may alter gene expression. Expression analyses identify genes that are differentially regulated by genotype (strain A fed diet 1 vs. strain B fed diet 1, and strain A vs. strain B fed diet 2) and diet (diet 1 in strain A vs. diet 2 in strain A, and diet 1 in

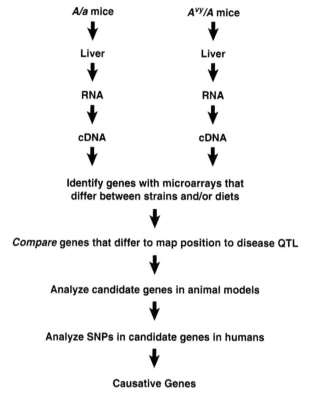

Figure 3.1   Experimental paradigm for identifying causal disease susceptibility genes. *A/a* and
Avy/A mice were fed micronutrient-balanced diets containing 70 or 100% calories
for 90 days (for details see J. Kaput et al., *Physiol Genomics*, 2004. 18(3): p.
316–24).

strain B vs. diet 2 in strain B). Genes that are regulated differently, depending upon
diet and genetic makeup (for example, by high-fat diet in strain A but low-fat diet
in strain B), are genes regulated by genotype X diet interactions. Such genotype X
diet regulated genes may contribute to the complexity of identifying genes involved
in chronic diseases.

Food intake is rigorously controlled in our studies so the nutrient status (fed vs.
fasted) is known.[67,92] Reducing biological and environmental variables is critical for
interpreting gene expression data and testing for reproducibility. Dietary supplements
or phytochemicals can also be analyzed with our experimental design by spiking
control diets with mixtures or purified chemicals.

In our most recent work,[43] groups of agouti mice (*A/a*) and obese yellow mice
(*Avy/A*) were fed either control diets (100% calories) or 70% calories. These strains
are genetically identical except for one locus: *Avy*, a mutation in the agouti gene,
which causes it to be over-expressed and expressed in multiple tissues. Agouti plays
a role in the melanocortin pathway involved in weight control.[98] The yellow (*Avy/A*)
mouse becomes obese (65 grams) within about 8 months, while the *A/a* mouse
weighs only ~20–25 grams at that age, which is typical of most mouse strains.[103]

The over-expression of the agouti protein also causes symptoms of diabetes (hyper-insulinemia, hyperglycemia) and increases the susceptibility to cancers in a number of tissues.[104] Micronutrients, fiber, and vitamins were adjusted to maintain similar intakes in both diets.

mRNA abundance (gene expression profiles) were analyzed in liver mRNA of *A/a* (agouti) and obese yellow mice (*A^{vy}/A*) fed 70% or 100% calories for 90 days. Genes that differed in regulation helped produce the differences between the two genotypes on each of the diets. However, genes that are differentially expressed by diet, genotype, or their interactions do not identify cause from effect genes: diet and genotype may regulate many genes but only a subset may play a role in initiating changes that lead from health to disease.

To identify cause from effect genes, we mapped genes that are differentially regulated to chromosomal regions that others have identified as contributing to complex diseases (Figure 3.1). These regions are typically determined by QTL analyses and are public-domain data derived by others.

One or more genes at some QTLs are likely to be regulated by diet, since unbalanced diets are known to increase the risk and severity of chronic diseases such as obesity and T2DM. Differentially regulated genes that map within QTL are candidates for contributing to the specific quantitative trait when the diet alters expression of the gene. The range of susceptibility a person inherits depends upon the particular combination of genes that are inherited, but the specific susceptibility within that range depends on how diet and other environmental variables affect the expression of the genetic information, specifically the genes within the QTL. Hence, disease susceptibility is not wholly deterministic, since a controllable variable, in this case diet, alters the expression of genetic information and the probability of developing the disease.

Our experimental system (Figure 3.1) identified 28 genes that were regulated by caloric intake, genotype, or diet X genotype and that mapped to QTL associated with T2DM or subphenotypes of T2DM (Table 3.1). A complete description of these results is in Kaput et al.[43] These genes were structural, enzymatic, and regulatory and are candidates for analyses in humans (see Figure 3.1). Eight of the murine QTL in Table 3.1 had analogous T2DM QTL in humans[25] and are marked in Table 3.1 by footnote d. An additional 59 genes were regulated by diet, genotype, and genotype X diet and mapped to QTL contributing to obesity (not shown). Additional research will be required to test the involvement of these genes in T2DM and obesity in mice and humans.

INDIVIDUALIZED NUTRITION

Personalized diets, or diets for groups of individuals with similar metabolic makeup, will require genetic testing. These tests will be developed by experiments described here, followed by studies in humans. Genetic tests will likely be done in conjunction with physicians and genetic and nutritional counselors to develop an individualized dietary plan based upon a personal genetic analysis and health status. Once the knowledge of gene–diet–disease is developed, these tests may be done soon after

Table 3.1 Diet, Genotype, and Diet x Genotype Regulated Genes at "Diabesity" QTL

QTL[a]	MGI_ID[b]	Name[c]	MGI_ID	Association with T2DM	Ref
Dbsty1	2149843	Isocitrate dehydrogenase 1 (NADP+)	96413	Enzyme activity not changed in 32 patients with diabetes relative to controls	[1]
Insq2[d]	1932506				
Nidd5	2154986	Never in mitosis-related expressed kinase 6	1339708	β cell mass decreased in Type 2 diabetes, Nek6 regulates initiation of mitosis	[82], [83]
Insq3	1932507	Hermansky-Pudlak syndrome 3	2153839	HSPs involved in lysosomal production; dysfunction of islet lysosomal system impairs glucose stimulated release	[21], [85]
Dbsty2	2149844	PDGF-α receptor	486984	Interacts with hedgehog, involved in insulin production in pancreas	[97]
Igfbp3q1	1890481	Shroom	237964	Regulates cytoarchitecture, adherens, and actin dynamics affect glucose uptake	[34], [49]
Fglu	2149370	Smoothened homolog	108075	Receptor for Indian Hedgehog (IHH), involved in insulin signaling in pancreas; Smo expressed in liver	[46], [97]
		Filamin	95557	Insulin causes changes in cytoskeleton, filamin A may interact with insulin receptor	[32]
Nidd3n[d]	1355301	Mitotic arrest deficient (homology)-like	186037	Component of mitotic spindle assembly checkpoint; may bind insulin receptor	[65]
		Erato Doi 32	243583	C2 domain — Calcium lipid binding domain IPR000008; calcium is important for insulin release (and other functions)	e.g., [55]
Igftbp3q2[d]	1890485	Amino terminal enhancer of split	88257	Co-repressor of NFκβ. NFκβ is activated in insulin-resistant tissues	[94], [87]
		Tumor rejection antigen	700006	Chaperonin regulated by glucose, involved in innate and specific immunity, c	[35]
Insq4[d]	1932516	Double minute 2	358044	May protect B-cell from fatty acid induced apoptosis; interacts with P53, involved in insulin receptor 3 regulation	[86, 105]
Nidd4n	1355273	Debrin	97919	Debrin = mABP1 = SH3P7, a target of Scr tyrosine kinase;implicated in cytoskeletal regulation, endocytosis, cAMP signaling	[48], [28]
		Disheveled 2	98553	Involved in complexes regulating Wnt pathway; Wnt inhibits glycogen synthase kinase-3 stabilizing β-catenin. GSK3 involved in T2DM?	[84]
Nidd4n	1355320	Nucleolar protein family A Member 2	545881	Indirect? Member of a complex involved in nucleolar RNA processing	[33]
		Septin 8	894310	Cell division and chromosome partitioning; no direct studies	
		S-phase kinase-associated protein	479801	In Arabidopsis, interacts with AtGRH1, a homolog of yeast GRR1, involved in glucose repressio	[95]
		Mannose-P-Dolichol utilization defect	497018	Helix-loop-helix transcription factor involved in pancreatic development and muscle function	[61], [3]

(continued)

Table 3.1　Diet, Genotype, and Diet x Genotype Regulated Genes at "Diabesity" QTL (continued)

QTL[a]	MGI_ID[b]	Name[c]	MGI_ID	Association with T2DM	Ref
Nidd4	68	Fatty acid synthase	351529	Expression regulated by insulin and glucose	[91], [23]
		Growth factor receptor bound protein 2	409816	Increased association of IRS-1/phosphatidylinositol 3-kinase, IRS-1/growth factor receptor bound 2 (Grb2), and Shc/Grb2 in diabetic rats	[96]
Dbsty3	2149845	Homeobox protein goosecoid	107717	None — Gsc involved in development	
Nidd2n	1355273	T-cell receptor alpha	98553	No information; immune functions important in diabetes	
Dbsty4	2149846	Glutamic pyruvic ttransaminase	95802	Diabetics are magnesium deficient; alterations in signal transduction? Cleaves PO_4 from TAK1, involved in inflammation	[16], [31], [59]
		Uncharacterized	106313		
		Thyrotroph embryonic factor, transcript variant	523917	In thymus, involved in calcium responsive genes expression	[51]
		Histone H1	523960	H1 involved in chromatin structure	
Insq5[d]	1932517	Protein phosphatase 1B, Mg^{2+}, β	523448	PTP1B as a negative regulator of insulin action	[77]
Nidd2[d]	1227794	Fibroblast growth factor	534640	Involved in insulin secretion in pancreas	[76]

a QTL loci description, numbers refer to specific loci

　　Dbsty　　　diabesity
　　Insq　　　insulin QTL
　　Nidd　　　non-insulin-dependent diabetes mellitus
　　Niddn　　　non-insulin-dependent diabetes mellitus in NSY
　　T2dm　　　type 2 diabetes mellitus
　　Igfbp3　　　insulin-like growth factor binding protein

b MGI_ID = Mouse Genome Identifier

c Name = enzyme

d Human T2DM QTL are syntenic to these QTL [25]

Source: From Kaput, J. et al., Physiol. Genomics, 18(3):316–324, 2004.

birth because long-term health may be dependent on developmental windows that may be affected by diet. However, society will have to debate and accept such testing and the resultant preventative approaches to health. National and international genetic privacy laws must be enacted before such testing becomes widely available. Society will also have to address a host of consequences of such genetic testing that include, for example, the possibility of offering insurance for different susceptibilities to disease, or penalties for individuals who do not follow nutritional advice based upon their genotypes. Although these ethical dilemmas are not currently relevant, such issues become real as these tests emerge from academic and clinical research.

Food companies may develop new markets, and novel foods are likely to evolve in tandem with the ability to identify genotypes.[22,26] Although it is unlikely that each individual will have a unique diet because of manufacturing issues, foods are likely to be designed for groups of individuals with similar genotypes. Individuals in these groups will come from a variety of ancestral populations, since genetic variation within an ethnic group is larger than between groups.[80] Specific nutrients such as vitamins, dietary supplements such as herbs, and minerals may be tailored to each individual based upon genetic testing of the relevant gene sets. One of the key challenges for implementing diets based on genotype will be analyzing the amount of a nutrient or bioactive each individual should consume — the field currently is examining whether a person can be helped by a nutrient, not how much. The outcome of this research is expected to profoundly influence human health and dietary habits; personalized diets are likely to maintain health and delay the effects of aging and aging-related disorders.

The study of chronic diseases such as obesity, cardiovascular diseases, and type 2 diabetes mellitus are already being impacted by nutritional genomics concepts and technologies.[42] These diseases are associated with food abundance and high caloric intake in affluent, industrialized countries. Since a large proportion of the world's population still suffers from malnutrition, the emphasis on studies of these diseases has been criticized for its focus on wealthy nations where calories and nutrients are in surplus. However, nutritional genomics research strategies and methods often involve comparative analyses — differences between two or more genotypes and two or more diets. Hence, nutrigenomic studies conducted in collaboration with individuals in developing countries are socially and scientifically desirable. The results and information from this field of research will have direct application for improving the health of the large numbers of people who lack adequate nutrition as well as those who over-consume.

ACKNOWLEDGMENTS

This work was supported by the National Center for Minority Health and Health Disparities Center of Excellence in Nutritional Genomics Grant MD-00222 and NutraGenomics (Chicago, IL).

REFERENCES

1. Belfiore, F, Romeo, F, Napoli, E, and Lo Vecchio, L, Enzymes of glucose metabolism in liver of subjects with adult-onset diabetes, *Diabetes*, 1974. **23**(4): 293–301.
2. Benbow, U, Tower, GB, Wyatt, CA, Buttice, G, and Brinckerhoff, CE, High levels of mmp-1 expression in the absence of the 2g single nucleotide polymorphism is mediated by p38 and erk1/2 mitogen-activated protein kinases in vmm5 melanoma cells, *J Cell Biochem*, 2002. **86**(2): 307–319.
3. Beranger, F, Mejean, C, Moniot, B, Berta, P, and Vandromme, M, Muscle differentiation is antagonized by sox15, a new member of the sox protein family, *J Biol Chem*, 2000. **275**(21): 16103–16109.
4. Black, SA, Diabetes, diversity, and disparity: what do we do with the evidence? *Am J Public Health*, 2002. **92**(4): 543–548.
5. Boden, G, Pathogenesis of type 2 diabetes. Insulin resistance, *Endocrinol Metab Clin North Am*, 2001. **30**(4): 801–815, v.
6. Bream, JH, Ping, A, Zhang, X, Winkler, C, and Young, HA, A single nucleotide polymorphism in the proximal ifn-gamma promoter alters control of gene transcription, *Genes Immun*, 2002. **3**(3): 165–169.
7. Brockmann, GA and Bevova, MR, Using mouse models to dissect the genetics of obesity, *Trends Genet*, 2002. **18**(7): 367–376.
8. Cao, SX, Dhahbi, JM, Mote, PL, and Spindler, SR, Genomic profiling of short- and long-term caloric restriction effects in the liver of aging mice, *Proc Natl Acad Sci USA*, 2001. **98**(19): 10630–10635.
9. Cardon, LR and Bell, JI, Association study designs for complex diseases, *Nat Rev Genet*, 2001. **2**(2): 91–99.
10. Clarke, SD and Abraham, S, Gene expression: Nutrient control of pre- and posttranscriptional events, *Faseb J*, 1992. **6**(13): 3146–3152.
11. Collins-Schramm, HE, Phillips, CM, Operario, DJ, Lee, JS, Weber, JL, Hanson, RL, Knowler, WC, Cooper, R, Li, H, and Seldin, MF, Ethnic-difference markers for use in mapping by admixture linkage disequilibrium, *Am J Hum Genet*, 2002. **70**(3): 737–750.
12. Cooney, CA, Dave, AA, and Wolff, GL, Maternal methyl supplements in mice affect epigenetic variation and DNA methylation of offspring, *J Nutr*, 2002. **132**(8 Suppl): 2393S-2400S.
13. Cousins, RJ, Nutritional regulation of gene expression, *Am J Med*, 1999. **106**(1A): 20S–23S; discussion 50S–51S.
14. Da Silva, P, Aitken, RP, Rhind, SM, Racey, PA, and Wallace, JM, Impact of maternal nutrition during pregnancy on pituitary gonadotrophin gene expression and ovarian development in growth-restricted and normally grown late gestation sheep fetuses, *Reproduction*, 2002. **123**(6): 769–777.
15. De Caterina, R, Madonna, R, Hassan, J, and Procopio, AD, Nutrients and gene expression, *World Rev Nutr Diet*, 2001. **89**: 23–52.
16. de Valk, HW, Magnesium in diabetes mellitus, *Neth J Med*, 1999. **54**(4): 139–146.
17. Deng, HW, Chen, WM, and Recker, RR, Population admixture: detection by Hardy-Weinberg test and its quantitative effects on linkage-disequilibrium methods for localizing genes underlying complex trait, *Genetics*, 2001. **157**(2): 885–897.
18. Elliott, TS, Swartz, DA, Paisley, EA, Mangian, HJ, Visek, WJ, and Kaput, J, F1fo-atpase subunit e gene isolated in a screen for diet regulated genes, *Biochem Biophys Res Commun*, 1993. **190**(1): 167–174.

19. Enattah, NS, Sahi, T, Savilahti, E, Terwilliger, JD, Peltonen, L, and Jarvela, I, Identification of a variant associated with adult-type hypolactasia, *Nat Genet,* 2002. **30**(2): 233–237.

20. Fafournoux, P, Bruhat, A, and Jousse, C, Amino acid regulation of gene expression, *Biochem J,* 2000. **351**(Pt 1): 1–12.

21. Feng, L, Novak, EK, Hartnell, LM, Bonifacino, JS, Collinson, LM, and Swank, RT, The Hermansky-Pudlak syndrome 1 (hps1) and hps2 genes independently contribute to the production and function of platelet dense granules, melanosomes, and lysosomes, *Blood,* 2002. **99**(5): 1651–1658.

22. Ferguson, LR and Kaput, J, Nutrigenomics and the New Zealand food industry, *Food New Zealand* (the journal of the New Zealand Institute of Food Science and Technology), 2004: 29–36.

23. Ferre, P, Regulation of gene expression by glucose, *Proc Nutr Soc,* 1999. **58**(3): 621–623.

24. Flint, J and Mott, R, Finding the molecular basis of quantitative traits: Successes and pitfalls, *Nat Rev Genet,* 2001. **2**(6): 437–445.

25. Florez, JC, Hirschhorn, J, and Altshuler, D, The inherited basis of diabetes mellitus: implications for the genetic analysis of complex traits, *Annu Rev Genomics Hum Genet,* 2003. **4**: 257–291.

26. Fogg-Johnson, N and Kaput, J, Nutrigenomics: an emerging scientific discipline, *Food Technology,* 2003. **57**(4): 61–67.

27. Forman, BM, Tontonoz, P, Chen, J, Brun, RP, Spiegelman, BM, and Evans, RM, 15-deoxy-delta 12, 14-prostaglandin j2 is a ligand for the adipocyte determination factor ppar gamma, *Cell,* 1995. **83**(5): 803–812.

28. Fucini, RV, Chen, JL, Sharma, C, Kessels, MM, and Stamnes, M, Golgi vesicle proteins are linked to the assembly of an actin complex defined by mabp1, *Mol Biol Cell,* 2002. **13**(2): 621–631.

29. Goodridge, AG, Dietary regulation of gene expression: enzymes involved in carbohydrate and lipid metabolism, *Annu Rev Nutr,* 1987. **7**: 157–185.

30. Goodridge, AG, The role of nutrients in gene expression, *World Rev Nutr Diet,* 1990. **63**: 183–193.

31. Hanada, M, Ninomiya-Tsuji, J, Komaki, K, Ohnishi, M, Katsura, K, Kanamaru, R, Matsumoto, K, and Tamura, S, Regulation of the tak1 signaling pathway by protein phosphatase 2c, *J Biol Chem,* 2001. **276**(8): 5753–5759.

32. He, HJ, Kole, S, Kwon, YK, Crow, MT, and Bernier, M, Interaction of filamin a with the insulin receptor alters insulin-dependent activation of the mitogen-activated protein kinase pathway, *J Biol Chem,* 2003. **278**(29): 27096–27104.

33. Henras, A, Henry, Y, Bousquet-Antonelli, C, Noaillac-Depeyre, J, Gelugne, JP, and Caizergues-Ferrer, M, Nhp2p and nop10p are essential for the function of h/aca snornps, *Embo J,* 1998. **17**(23): 7078–7090.

34. Hildebrand, JD and Soriano, P, Shroom, a pdz domain-containing actin-binding protein, is required for neural tube morphogenesis in mice, *Cell,* 1999. **99**(5): 485–497.

35. Hilf, N, Singh-Jasuja, H, and Schild, H, The heat shock protein gp96 links innate and specific immunity, *Int J Hyperthermia,* 2002. **18**(6): 521–533.

36. Hirschhorn, JN, Lohmueller, K, Byrne, E, and Hirschhorn, K, A comprehensive review of genetic association studies, *Genet Med,* 2002. **4**(2): 45–61.

37. Hollox, EJ, Poulter, M, Wang, Y, Krause, A, and Swallow, DM, Common polymorphism in a highly variable region upstream of the human lactase gene affects DNA-protein interactions, *Eur J Hum Genet,* 1999. **7**(7): 791–800.

38. Huang, Z, Bodkin, NL, Ortmeyer, HK, Zenilman, ME, Webster, NJ, Hansen, BC, and Shuldiner, AR, Altered insulin receptor messenger ribonucleic acid splicing in liver is associated with deterioration of glucose tolerance in the spontaneously obese and diabetic rhesus monkey: analysis of controversy between monkey and human studies, *J Clin Endocrinol Metab,* 1996. **81**(4): 1552–1556.

39. Itoh, Y, Kawamata, Y, Harada, M, Kobayashi, M, Fujii, R, Fukusumi, S, Ogi, K, Hosoya, M, Tanaka, Y, Uejima, H, Tanaka, H, Maruyama, M, Satoh, R, Okubo, S, Kizawa, H, Komatsu, H, Matsumura, F, Noguchi, Y, Shinohara, T, Hinuma, S, Fujisawa, Y, and Fujino, M, Free fatty acids regulate insulin secretion from pancreatic beta cells through gpr40, *Nature,* 2003. **422**(6928): 173–176.

40. Jackson, AA, Nutrients, growth, and the development of programmed metabolic function, *Adv Exp Med Biol,* 2000. **478**: 41–55.

41. Jimenez-Sanchez, G, Childs, B, and Valle, D, Human disease genes, *Nature,* 2001. **409**(6822): 853–855.

42. Kaput, J, Diet–disease gene interactions, *Nutrition,* 2004. **20**(1): 26–31.

43. Kaput, J, Klein, KG, Reyes, EJ, Kibbe, WA, Cooney, CA, Jovanovic, B, Visek, WJ, and Wolff, GL, Identification of genes contributing to the obese yellow avy phenotype: caloric restriction, genotype, diet x genotype interactions, *Physiol Genomics,* 2004. **18**(3): 316–324.

44. Kaput, J and Rodriguez, RL, Nutritional genomics: the next frontier in the postgenomic era, *Physiol Genomics,* 2004. **16**(2): 166–177.

45. Kaput, J, Swartz, D, Paisley, E, Mangian, H, Daniel, WL, and Visek, WJ, Diet–disease interactions at the molecular level: an experimental paradigm. *J Nutr,* 1994. **124**(8 Suppl): 1296S–1305S.

46. Kayed, H, Kleeff, J, Keleg, S, Buchler, MW, and Friess, H, Distribution of Indian hedgehog and its receptors patched and smoothened in human chronic pancreatitis, *J Endocrinol,* 2003. **178**(3): 467–478.

47. Keita, SO, Kittles, RA, Royal, CD, Bonney, GE, Furbert-Harris, P, Dunston, GM, and Rotimi, CN, Conceptualizing human variation, *Nat Genet,* 2004. **36 Suppl 1**: S17–20.

48. Kessels, MM, Engqvist-Goldstein, AE, and Drubin, DG, Association of mouse actin-binding protein 1 (mabp1/sh3p7), an src kinase target, with dynamic regions of the cortical actin cytoskeleton in response to rac1 activation, *Mol Biol Cell,* 2000. **11**(1): 393–412.

49. Khan, AH and Pessin, JE, Insulin regulation of glucose uptake: a complex interplay of intracellular signalling pathways, *Diabetologia,* 2002. **45**(11): 1475–1483.

50. Kliewer, SA, Lenhard, JM, Willson, TM, Patel, I, Morris, DC, and Lehmann, JM, A prostaglandin j2 metabolite binds peroxisome proliferator-activated receptor gamma and promotes adipocyte differentiation, *Cell,* 1995. **83**(5): 813–819.

51. Krueger, DA, Warner, EA, and Dowd, DR, Involvement of thyrotroph embryonic factor in calcium-mediated regulation of gene expression, *J Biol Chem,* 2000. **275**(19): 14524–14531.

52. Kuiper, GG, Carlsson, B, Grandien, K, Enmark, E, Haggblad, J, Nilsson, S, and Gustafsson, JA, Comparison of the ligand binding specificity and transcript tissue distribution of estrogen receptors alpha and beta, *Endocrinology,* 1997. **138**(3): 863–870.

53. Lamba, JK, Lin, YS, Schuetz, EG, and Thummel, KE, Genetic contribution to variable human cyp3a-mediated metabolism, *Adv Drug Deliv Rev,* 2002. **54**(10): 1271–1294.

54. Lander, E and Kruglyak, L, Genetic dissection of complex traits: guidelines for interpreting and reporting linkage results, *Nat Genet,* 1995. **11**(3): 241–247.

55. Lang, J, Fukuda, M, Zhang, H, Mikoshiba, K, and Wollheim, CB, The first c2 domain of synaptotagmin is required for exocytosis of insulin from pancreatic beta-cells: action of synaptotagmin at low micromolar calcium, *Embo J,* 1997. **16**(19): 5837–5846.

56. Lee, C, Atanelov, L, Modrek, B, and Xing, Y, Asap: the alternative splicing annotation project, *Nucleic Acids Res,* 2003. **31**(1): 101–105.

57. Lee, CK, Allison, DB, Brand, J, Weindruch, R, and Prolla, TA, Transcriptional profiles associated with aging and middle age-onset caloric restriction in mouse hearts, *Proc Natl Acad Sci USA,* 2002. **99**(23): 14988–14993.

58. Lee, CK, Klopp, RG, Weindruch, R, and Prolla, TA, Gene expression profile of aging and its retardation by caloric restriction, *Science,* 1999. **285**(5432): 1390–1393.

59. Lee, J, Mira-Arbibe, L, and Ulevitch, RJ, Tak1 regulates multiple protein kinase cascades activated by bacterial lipopolysaccharide, *J Leukoc Biol,* 2000. **68**(6): 909–915.

60. Linder, CC, The influence of genetic background on spontaneous and genetically engineered mouse models of complex diseases, *Lab Anim (NY),* 2001. **30**(5): 34–39.

61. Lioubinski, O, Muller, M, Wegner, M, and Sander, M, Expression of sox transcription factors in the developing mouse pancreas, *Dev Dyn,* 2003. **227**(3): 402–408.

62. McKeigue, PM, Carpenter, JR, Parra, EJ, and Shriver, MD, Estimation of admixture and detection of linkage in admixed populations by a bayesian approach: application to African-American populations, *Ann Hum Genet,* 2000. **64**(Pt 2): 171–186.

63. Morris, SM, Jr., Nilson, JH, Jenik, RA, Winberry, LK, McDevitt, MA, and Goodridge, AG, Molecular cloning of gene sequences for avian fatty acid synthase and evidence for nutritional regulation of fatty acid synthase mrna concentration, *J Biol Chem,* 1982. **257**(6): 3225–3229.

64. Nathan, DM, Clinical practice. initial management of glycemia in type 2 diabetes mellitus, *N Engl J Med,* 2002. **347**(17): 1342–1349.

65. O'Neill, TJ, Zhu, Y, and Gustafson, TA, Interaction of mad2 with the carboxyl terminus of the insulin receptor but not with the igfir. Evidence for release from the insulin receptor after activation, *J Biol Chem,* 1997. **272**(15): 10035–10040.

66. Pagel, M and Mace, R, The cultural wealth of nations, *Nature,* 2004. **428**(6980): 275–278.

67. Paisley, EA, Park, EI, Swartz, DA, Mangian, HJ, Visek, WJ, and Kaput, J, Temporal-regulation of serum lipids and stearoyl coa desaturase and lipoprotein lipase mrna in balb/chnn mice, *J Nutr,* 1996. **126**(11): 2730–2737.

68. Park, EI, Paisley, EA, Mangian, HJ, Swartz, DA, Wu, MX, O'Morchoe, PJ, Behr, SR, Visek, WJ, and Kaput, J, Lipid level and type alter stearoyl coa desaturase mrna abundance differently in mice with distinct susceptibilities to diet-influenced diseases, *J Nutr,* 1997. **127**(4): 566–573.

69. Parra, EJ, Marcini, A, Akey, J, Martinson, J, Batzer, MA, Cooper, R, Forrester, T, Allison, DB, Deka, R, Ferrell, RE, and Shriver, MD, Estimating African American admixture proportions by use of population-specific alleles, *Am J Hum Genet,* 1998. **63**(6): 1839–1851.

70. Paterson, A, Damon S, Hewitt JD, Zamir D, Rabinowitch HD, Lincoln SE, Lander ES, Tanksley SD, Mendelian factors underlying quantitative traits in tomato: comparison across species, generations, and environments, *Genetics,* 1991. **127**(1): 181–197.

71. Pennacchio, LA, Insights from human/mouse genome comparisons, *Mamm Genome,* 2003. **14**(7): 429–436.

72. Pfaff, CL, Parra, EJ, Bonilla, C, Hiester, K, McKeigue, PM, Kamboh, MI, Hutchinson, RG, Ferrell, RE, Boerwinkle, E, and Shriver, MD, Population structure in admixed populations: effect of admixture dynamics on the pattern of linkage disequilibrium. *Am J Hum Genet,* 2001. **68**(1): 198–207.

73. Pianetti, S, Guo, S, Kavanagh, KT, and Sonenshein, GE, Green tea polyphenol epigallocatechin-3 gallate inhibits her-2/neu signaling, proliferation, and transformed phenotype of breast cancer cells, *Cancer Res,* 2002. **62**(3): 652–655.

74. Prolla, TA, DNA microarray analysis of the aging brain, *Chem Senses,* 2002. **27**(3): 299–306.

75. Radich, JP, Mao, M, Stepaniants, S, Biery, M, Castle, J, Ward, T, Schimmack, G, Kobayashi, S, Carleton, M, Lampe, J, and Linsley, PS, Individual-specific variation of gene expression in peripheral blood leukocytes, *Genomics,* 2004. **83**(6): 980–988.

76. Raile, K, Berthold, A, Banning, U, Horn, F, Pfeiffer, G, and Kiess, W, Igfs, basic fgf, and glucose modulate proliferation and apoptosis induced by ifngamma but not by il-1beta in rat ins-1e beta-cells, *Horm Metab Res,* 2003. **35**(7): 407–414.

77. Ramachandran, C and Kennedy, BP, Protein tyrosine phosphatase 1b: a novel target for type 2 diabetes and obesity, *Curr Top Med Chem,* 2003. **3**(7): 749–757.

78. Reich, DE and Goldstein, DB, Detecting association in a case-control study while correcting for population stratification, *Genet Epidemiol,* 2001. **20**(1): 4–16.

79. Risch, N, Evolving methods in genetic epidemiology. Ii. Genetic linkage from an epidemiologic perspective, *Epidemiol Rev,* 1997. **19**(1): 24–32.

80. Risch, N, Burchard, E, Ziv, E, and Tang, H, Categorization of humans in biomedical research: Genes, race and disease, *Genome Biol,* 2002. **3**(7): comment 2007.

81. Risch, N, Ghosh, S, and Todd, JA, Statistical evaluation of multiple-locus linkage data in experimental species and its relevance to human studies: application to non-obese diabetic (nod) mouse and human insulin-dependent diabetes mellitus (iddm), *Am J Hum Genet,* 1993. **53**(3): 702–714.

82. Ritzel, RA and Butler, PC, Replication increases beta-cell vulnerability to human islet amyloid polypeptide-induced apoptosis, *Diabetes,* 2003. **52**(7): 1701–1708.

83. Roig, J, Mikhailov, A, Belham, C, and Avruch, J, Nercc1, a mammalian nima-family kinase, binds the ran gtpase and regulates mitotic progression, *Genes Dev,* 2002. **16**(13): 1640–1658.

84. Sakanaka, C, Sun, TQ, and Williams, LT, New steps in the wnt/beta-catenin signal transduction pathway, *Recent Prog Horm Res,* 2000. **55**: 225–236.

85. Salehi, A, Henningsson, R, Mosen, H, Ostenson, CG, Efendic, S, and Lundquist, I, Dysfunction of the islet lysosomal system conveys impairment of glucose-induced insulin release in the diabetic gk rat, *Endocrinology,* 1999. **140**(7): 3045–3053.

86. Sciacchitano, S, Orecchio, A, Lavra, L, Misiti, S, Giacchini, A, Zani, M, Danese, D, Gurtner, A, Soddu, S, Di Mario, U, and Andreoli, M, Cloning of the mouse insulin receptor substrate-3 (mirs-3) promoter, and its regulation by p53, *Mol Endocrinol,* 2002. **16**(7): 1577–1589.

87. Shoelson, SE, Lee, J, and Yuan, M, Inflammation and the ikk beta/i kappa b/nf-kappa b axis in obesity- and diet-induced insulin resistance, *Int J Obes Relat Metab Disord,* 2003. **27 Suppl 3**: S49–52.

88. Sing, CF, Stengard, JH, and Kardia, SL, Genes, environment, and cardiovascular disease, *Arterioscler Thromb Vasc Biol,* 2003. 1190–1196.

89. Stoehr, JP, Nadler, ST, Schueler, KL, Rabaglia, ME, Yandell, BS, Metz, SA, and Attie, AD, Genetic obesity unmasks nonlinear interactions between murine type 2 diabetes susceptibility loci, *Diabetes,* 2000. **49**(11): 1946–1954.

90. Sugimoto, K, Murakawa, Y, Zhang, W, Xu, G, and Sima, AA, Insulin receptor in rat peripheral nerve: its localization and alternatively spliced isoforms, *Diabetes Metab Res Rev,* 2000. **16**(5): 354–363.

91. Sul, HS, Latasa, MJ, Moon, Y, and Kim, KH, Regulation of the fatty acid synthase promoter by insulin, *J Nutr,* 2000. **130**(2S Suppl): 315S–320S.

92. Swartz, DA, Park, EI, Visek, WJ, and Kaput, J, The e subunit gene of murine f1f0-atp synthase. Genomic sequence, chromosomal mapping, and diet regulation, *J Biol Chem,* 1996. **271**(34): 20942–20948.

93. Tabor, HK, Risch, NJ, and Myers, RM, Opinion: candidate-gene approaches for studying complex genetic traits: practical considerations, *Nat Rev Genet,* 2002. **3**(5): 391–397.

94. Tetsuka, T, Uranishi, H, Imai, H, Ono, T, Sonta, S, Takahashi, N, Asamitsu, K, and Okamoto, T, Inhibition of nuclear factor-kappab-mediated transcription by association with the amino-terminal enhancer of split, a groucho-related protein lacking wd40 repeats, *J Biol Chem,* 2000. **275**(6): 4383–4390.

95. Thelander, M, Fredriksson, D, Schouten, J, Hoge, JH, and Ronne, H, Cloning by pathway activation in yeast: identification of an arabidopsis thaliana f-box protein that can turn on glucose repression, *Plant Mol Biol,* 2002. **49**(1): 69–79.

96. Thirone, AC, Scarlett, JA, Gasparetti, AL, Araujo, EP, Lima, MH, Carvalho, CR, Velloso, LA, and Saad, MJ, Modulation of growth hormone signal transduction in kidneys of streptozotocin-induced diabetic animals: effect of a growth hormone receptor antagonist, *Diabetes,* 2002. **51**(7): 2270–2281.

97. Thomas, MK, Rastalsky, N, Lee, JH, and Habener, JF, Hedgehog signaling regulation of insulin production by pancreatic beta-cells, *Diabetes,* 2000. **49**(12): 2039–2047.

98. Voisey, J and van Daal, A, Agouti: from mouse to man, from skin to fat, *Pigment Cell Res,* 2002. **15**(1): 10–18.

99. Wentworth, JM, Agostini, M, Love, J, Schwabe, JW, and Chatterjee, VK, St John's wort, a herbal antidepressant, activates the steroid x receptor, *J Endocrinol,* 2000. **166**(3): R11–16.

100. Whitney, AR, Diehn, M, Popper, SJ, Alizadeh, AA, Boldrick, JC, Relman, DA, and Brown, PO, Individuality and variation in gene expression patterns in human blood, *Proc Natl Acad Sci USA,* 2003. **100**(4): 1896–1901.

101. Whorwood, CB, Firth, KM, Budge, H, and Symonds, ME, Maternal undernutrition during early to midgestation programs tissue-specific alterations in the expression of the glucocorticoid receptor, 11beta-hydroxysteroid dehydrogenase isoforms, and type 1 angiotensin ii receptor in neonatal sheep, *Endocrinology,* 2001. **142**(7): 2854–2864.

102. Williams, RJ, *Biochemical Individuality,* 1956, Austin, TX: John Wiley & Sons.

103. Wolff, GL, Kodell, RL, Kaput, JA, and Visek, WJ, Caloric restriction abolishes enhanced metabolic efficiency induced by ectopic agouti protein in yellow mice, *Proc Soc Exp Biol Med,* 1999. **221**(2): 99–104.

104. Wolff, GL, Roberts, DW, and Mountjoy, KG, Physiological consequences of ectopic agouti gene expression: the yellow obese mouse syndrome, *Physiol Genomics,* 1999. **1**(3): 151–163.

105. Wrede, CE, Dickson, LM, Lingohr, MK, Briaud, I, and Rhodes, CJ, Protein kinase b/akt prevents fatty acid-induced apoptosis in pancreatic beta-cells (ins-1), *J Biol Chem,* 2002. **277**(51): 49676–49684.

106. Yan, H, Yuan, W, Velculescu, VE, Vogelstein, B, and Kinzler, KW, Allelic variation in human gene expression, *Science,* 2002. **297**(5584): 1143.

CHAPTER 4

Anti-Inflammatory Phytochemicals: *In Vitro* and *Ex Vivo* Evaluation

Marc Lemay

CONTENTS

PERILS AND BENEFITS OF ANTI-INFLAMMATION

Aspirin, along with ibuprofen and other non-steroid anti-inflammatory drugs (NSAIDs), is associated with injury to the gastrointestinal (GI) mucosa. Chronic NSAID users, the elderly, and those who have had an ulcer in the past are particularly susceptible to NSAID-induced gastropathy;[1] but even short-term use can cause detectable gastric damage.[2] NSAIDs are believed to cause injury to the GI tract at least in part by the unselective inhibition of the enzyme cyclooxygenase (Cox). The Cox enzyme has two isoforms, Cox-1 and Cox-2. Cox-1 supports renal and platelet function and protects the GI mucosa, and Cox-2 mediates pain and inflammation. Existing NSAIDs inhibit Cox-1 more efficiently than they do Cox-2, and the pain-relieving Cox-2 inhibition they provide comes with the drawback of relatively greater Cox-1 inhibition. This can lead to mild gastric erosions, bleeding ulcers, and increased risk of anemia from occult fecal blood loss or catastrophic internal bleeding.[3] The pharmaceutical industry, in response to this real need, concentrated on developing a safer anti-inflammatory and analgesic drug for long-term use; the first one to be launched in the U.S. was celecoxib (Celebrex), followed by rofecoxib (Vioxx).

Intermittent, low-dose use of selective Cox-2 inhibitors such as celecoxib causes fewer gastrointestinal side effects than unselective drugs.[4-6] However, selective Cox-2 inhibition is available only on prescription, is cost-effective only for high-risk subjects,[7] and is not without its own particular risks. Recently, high-dose, long-term use of rofecoxib, a drug with about ten-fold the Cox-2 selectivity of celecoxib,[8] was reported to cause heart problems in some patients; in response, its manufacturer withdrew it from the market.[9,10]

The over-the-counter market in Cox-2 inhibition remains divided between aspirin and ibuprofen. These two are not equally unselective; *in vitro*, ibuprofen shows a modest differential in selectivity toward Cox-2 compared to aspirin.[8] The shift in emphasis of drug action is small, yet yields a clinically important gastric safety advantage for ibuprofen.[11]

That ibuprofen is safer than aspirin indicates a possible therapeutic window of opportunity with regard to anti-inflammatory phytochemicals. Although Cox-2 inhibition is associated in the popular mind with modern and expensive pharmaceutical drugs, it is actually one of mankind's oldest therapeutic drug targets. Plant-based, pain-relieving Cox-2 inhibitors of one form or another have been consumed for approximately 3,000 years.[12] White willow bark, the traditional pain-and-fever remedy,[13,14] is a rich source of salicin,[15] a molecule that was just over 100 years ago elaborated into the most popular pharmaceutical in the world today, aspirin. Aspirin is not a selective Cox-2 inhibitor, but given the ubiquity of low-concentration anti-inflammatory compounds in a healthy diet,[16] one would expect to find variations on this pharmaceutical theme in nature.[17] Perhaps among all the known and obscure anti-inflammatory phytochemical agents there are a few that possess a favorable Cox-2 selectivity ratio, somewhere between that of ibuprofen and rofecoxib.

Phytochemical anti-inflammation is also of interest for another reason. One thriving focus of the young field of nutrigenomics — understood as the elucidation of disease-associated genetic variations, followed by their identification in a person who is then offered a bespoke dietary recommendation to counter his specific genetic susceptibility — has been cancer prevention. And closely related to cancer is inflammation.[18] One of the most important mediators in the body's inflammatory response is Cox-2, both the gene and the protein. While the role of the Cox-2 enzyme in pain and inflammation is well known,[19,20] it is becoming increasingly apparent that many anti-inflammatory drugs previously understood as Cox-2 *enzyme* inhibitors also work by inhibiting the Cox-2 *gene*,[21,22] a finding that could be of capital importance in the treatment or prevention of cancer[23–25] and other aging-related diseases.[26,27] A plant-based anti-inflammatory that would work both by inhibiting the Cox-2 enzyme (such as ibuprofen and other NSAIDs, or willow bark extract) as well as by inhibiting the Cox-2 gene (such as theaflavin[28] or humulon from hops[29,30]) could potentially be used both for mild pain relief and cancer prevention.

SCREENING FOR ANTI-INFLAMMATORY PHYTOCHEMICALS

The general rule in dietary supplement research, no less than in drug research, is that cost and laboriousness of development efforts increase in tandem with the predictive utility of the information being generated — predictive, that is, of performance in what is probably the costliest and most laborious device in all of product development, the clinical trial. The goal is to be able to make a substantiated, health-related statement about the dietary supplement's effects in healthy humans. A clinical trial is not always necessary. The law holds that marketing language be true and not misleading; it does not require that it be original or striking. Many label claims can be substantiated with reference to the published literature. However, novel claims must have support built up around them, piece by piece; in one common approach, the first piece will be the *in vitro* bioassay. This is the quickest way to get an answer. The answer will not be definitive, it might even be misleading; but this step is simple and inexpensive enough that it is hard to justify skipping it. After the *in vitro* assay, if appropriate, comes the *ex vivo* bioassay. Here, a test article is given to volunteers, and blood or other biological samples are taken from them at intervals after consumption. The samples are then prepared and tested *in vitro* in various ways. This method has the signal advantage of providing a functional bioavailability measure. Given whatever result a first-pass *in vitro* assay produced, that result could only be replicated *ex vivo* if the product is actually absorbed and distributed to the circulation in an active form by humans. This model can still fail to predict clinical effectiveness, because the desired reaction or effect still takes place outside the body. The final step in this stepwise screening model, then, is the *in vivo* test, whether clinical trial, or other means, of directly measuring a desired change or reaction in the human body.

The first step is deciding on the relevant outcome measures. To be sure, nociception, which is the activity produced in the nervous system by noxious stimuli, is a complex psychophysiological phenomena involving many more mediators than Cox-2. There are pain-relieving or anti-inflammatory phytochemicals with a wide variety of mechanisms of action, postulated or verified;[31] and for drawing inferences about the possible pain-relieving or anti-inflammatory properties of phytochemicals, many other bioassays are available.[32] But many of those plants with "other" mechanisms of action are unmarketable, either for political reasons (cannabis, opium), consumer resistance (capsaicin, available as a nasal spray), or supply challenges due to chaos in the country of origin.[17] And none of the other possible outcome measures has the advantages that Cox-2 inhibition has, of being not only an activity widely found in natural substances,[16] and linked to a wide body of research about the health benefits of anti-inflammation; but also, on a technical level, Cox-2 inhibition assays have also been well-defined for both *in vitro*[8] and *ex vivo*[33] testing, with published data providing reasonably strong links to clinical efficacy benchmarks.[34]

The disadvantage of such a choice is the possibility of missing out on discovering an entirely novel mechanism of action, as happened with, for example, the dietary supplement *Sangre de grado*, an Amazonian tree sap which appears to inhibit neurogenic inflammation by a recently discovered physiological pathway.[35]

Botanical agents that have been shown to inhibit Cox-2 activity *in vitro* include holy basil,[36,37] curcumin,[38–42] ginger and related products,[43,44] green tea, hops, berberine, and willow bark.[45] However, in spite of extensive laboratory evidence that these botanical agents can inhibit Cox-2 *in vitro*, and promising results from clinical trials (e.g., of ginger and willow for pain[46]), it has not yet been demonstrated that oral consumption of botanical agents can inhibit Cox-2 activity in humans.

MATERIALS AND METHODS

In Vitro Study

Cyclooxygenase-1 and -2 Assay

The potential of various botanical materials to inhibit Cox-1 and Cox-2 activity was assessed in phorbol 12-myristate 13-acetate- (PMA) stimulated Caco-2 cells. Caco-2 cells were maintained in ATCC recommended media (minimum essential medium with 2 mM L-glutamate and Earle's BSS adjusted to contain 1.5 g/L sodium bicarbonate, 0.1 mM non-essential amino acids, and 1.0 mM sodium pyruvate, and 20% fetal bovine serum). Cells seeded into 96-well plates in growth media were pretreated in triplicates with varying concentrations of treatment samples (range: 0–1000 µg/ml) for 4 hours prior to stimulation with 50 ng/ml PMA for an additional 20 hours to increase PGE_2 secretion. Following stimulation with PMA, cells were incubated with 10 µM arachidonic acid for 1 hour. PGE_2 secreted into the culture medium was quantitated using an ELISA kit specific for PGE_2 (R&D Systems, Minneapolis, MN).

Botanicals

We identified putative Cox-2 inhibiting or anti-inflammatory botanicals from the literature, from ethnopharmaceutical reports, and from supplier-provided documentation. Samples were obtained from commercial suppliers. The identification of the samples was predicated on visual inspection and the accompanying product data sheets or certificates of analysis. The tested botanicals are listed in Table 4.1. For the *in vitro* study, botanical samples were mixed into a solution of 50% DMSO, 30% ethanol, and 20% distilled water to a concentration of 50–100 mg/ml. Three Cox-inhibiting drugs were also tested as positive controls: ibuprofen, aspirin, and celecoxib.

Cell Culture

A549, a human epithelial carcinoma cell line (ECACC Ref. No. 86012804), expresses Cox-2 when exposed to IL-1β.[47] Production of PGE_2 by this cell line can therefore be used as an index of Cox-2 activity. A549 cells were maintained in a humidified atmosphere of 5% CO_2, 95% air at 37°C, and grown in Dulbecco's Modified Eagle Medium (DMEM) supplemented with 10% fetal bovine serum (FBS). For the experimental procedures, cells were seeded into 96-well plates and grown to confluence before use. In order to induce Cox-2 expression, A549 cells were incubated for 24 h in fresh DMEM supplemented with 10% FBS and IL-1β at a concentration of 10 ng/ml. Before the experimental procedure, the medium was replaced with 50 µl/well of fresh DMEM: Ca^{2+}-free modified Krebs-Ringer solution (see below) (4:1, v/v) at 37°C.

Washed Platelets

The production of TXB_2 by platelets was used as an index of Cox-1 activity. Blood from healthy volunteers, who had not taken NSAIDs for at least two weeks, was collected by venipuncture into gelatin-coated (0.1% porcine gelatin in H_2O, 1–3 h at 37°C) plastic tubes containing trisodium citrate 3.15% (1:9, v/v). The blood was centrifuged at 200 g for 7 min to produce platelet-rich plasma (PRP). Prostacyclin (300 ng/ml) was then added to the PRP followed by centrifugation at 1000 g for 15 min to sediment the platelets. The resulting supernatant was removed and replaced with an equal volume of Ca^{2+}-free modified Krebs-Ringer solution at 37°C (10 mM HEPES, 20 mM $NaHCO_3$, 120 mM NaCl, 4 mM KCl, 2 mM Na_2SO_4, 0.1% glucose, 0.1% bovine serum albumin). The pellet was gently suspended and further prostacyclin (300 ng/ml) added. The platelets were centrifuged again and suspended in Ca^{2+}-free modified Krebs-Ringer buffer at 37°C to match one fourth of the initial plasma volume. Thirty minutes later the platelet suspension was diluted 1:5 in DMEM supplemented with 10% FBS and plated into gelatin-coated 96-well plates (50 µl/well).

Evaluation of NSAIDs Activity on Cox-1 and Cox-2

To assay NSAID activity in plasma collected from the volunteers, 50 µl of plasma was added to medium bathing either pre-induced A549 cells or washed platelets. After incubation for 30 min at 37°C, calcium ionophore A23187 (50 µM) was added and the cells or platelets were incubated for an additional 15 min at 37°C. At the

Table 4.1 In Vitro Cox Inhibition Results

Test Product	Scientific Name	Marker Compound	Tested Concentrations (μg/ml)	Cox-2 (PGE2 secretion)		Cox-1 (TXB2 secretion)	
				IC_{50} (μg/ml)	IC_{80} (μg/ml)	IC_{50} (μg/ml)	IC_{80} (μg/ml)
Advil	ibuprofen	ibuprofen	0,1,3,10,30,100	0.6–0.7	2–3	0.1–0.2	>100
Bayer Aspirin	acetylsalicylic acid	acetylsalicylic acid	0,10,30,100,300,1000	1–2	8–9	0.3–0.4	>100
Celebrex	celecoxib	celecoxib	0,1,3,10,30,100	0.05–0.06	0.8–0.9	3–4	>100
Grape Extract	Vitis Vinifera	grape skin, seeds, stem extract, 40% polyphenols	0,10,30,100,300,1000	>1000	>1000	1000	>1000
Resveratrol	Vitis Vinifera	grape stem extract, 20% resveratrol oligostilbenes	0,1,3,10,30,100	3–4	9–10	5–6	>100
Hops Extract	Humulus lupulus	30% alpha acids	0,10,30,100,300,1000	1–2	3–4	20–30	>100
Chinese Skullcap	Scutellaria baicalensis	dried root powder	0,0.3,1,3,10,30,100	20–30	>100	>100	>100
Grains of Paradise	Afromomum stipulatum	ethanolic extract of ground seeds	0,3,10,30,100,300,1000	<3	<3	5–6	>30
Habanero Chili	Capsicum frutescens	ground dried fruit	0,3,10,30,100,300,1000	70–80	>1000	400–500	>1000
Rosemary Extract	Rosmarinus officinalis	50% carnosic acid	0,3,10,30,100,300,1000	30–40	100–200	>1000	>1000
Turmeric	Curcuma longa	raw root powder	0,3,10,30,100,300,1000	<3	3–4	7–8	700–800
Curcumin	Curcuma longa	95% Curcuminoid	0,3,10,30,100,300,1000	2–3	10–20	10–20	800–900
Sage Extract	Salvia officinalis L	dried powder	0,3,10,30,100,300,1000	>1000	>1000	>1000	>1000
Cayenne	Capsicum annuum	dried powder	0,3,10,30,100,300,1000	200–300	>1000	>1000	>1000
Thai Dragon	Capsicum annuum var. annuum	dried powder	0,3,10,30,100,300,1000	300–400	>1000	>1000	>1000
Devil's Claw	Harpagophytum procumbens	5% harpagoside	0,3,10,30,100,300,1000	>1000	>1000	1000	>1000
Myrrh	Commiphora Myrrha	resin	0,3,10,30,100,300,1000	6–7	9–10	900–1000	>1000

ANTI-INFLAMMATORY PHYTOCHEMICALS: *IN VITRO* AND *EX VIVO* EVALUATION   47

Common name	Botanical	Preparation	Concentration				
Frankincense	Boswellia Carteri	resin	0,3,10,30,100,300,1000	30–40	80–90	200–300	>1000
Oregano	Origanum vulgare L.	dried powder	0,3,10,30,100,300,1000	6–7	20–30	90–100	>1000
Wild Blueberry	Vaccinnium augustifolium	20% Anthocyanidins	0,3,10,30,100,300,1000	<3	3	50–60	>1000

Note: Each botanical was evaluated at the indicated concentration in triplicate determinations. Average values at each concentrated were plotted against concentration and dose-response EC_{50} and EC_{80} estimates were obtained by extrapolation.

end of the incubation, plates containing the platelet suspension were centrifuged for 5 min at 1500 g (4°C) and the supernatant removed and snap frozen until analysis by radioimmunoassay. Medium from A549 plates was also removed and frozen.

Ex Vivo Cox Inhibition

This was a single-center, randomized, active-controlled, double-blind parallel-groups study performed at ABG. The clinical protocol was reviewed and approved by the Western Institutional Review Board (Olympia, WA), and the study was performed according to Good Clinical Practice standards. Each subject provided written informed consent.

Validation Study

Ex vivo Cox inhibition was measured by the method of Giuliano and colleagues,[34] who have demonstrated congruence between the *ex vivo* Cox-inhibitory potencies and selectivities of selected NSAIDs and their known pain-relieving and gastric tract-damaging (or -sparing) properties.[33] First, to validate the assay methods, blood from two healthy volunteers was collected before, as well as 30, 60, and 120 min after a single dose of either ibuprofen 400 mg or celecoxib 200 mg. Plasma containing NSAIDs and metabolites was evaluated for its effect on Cox-1 and Cox-2 activity *ex vivo* in calcium ionophore treated washed human platelets and A549 cells (see below for detail), respectively. Measuring TxB_2 and PGE_2 as outcomes, inhibition of Cox-1 and Cox-2 by NSAID controls was demonstrated by expressing the concentrations of TxB_2 and PGE_2 produced as a percentage of basal. Blood samples were split and sent for analysis both at Alticor Analytical Services laboratory (Ada, MI) and at William Harvey Research Limited (London, U.K.).

Clinical Ex Vivo Study

Subjects

Nineteen, healthy adult volunteers, with a mean age (± SD) of 42.3 ± 12.6 years (range 24–65), participated in this study. Exclusion criteria included ulcer disease or history of any bleeding from the GI tract, uncontrolled hypertension, use of anti-inflammatory or analgesic drugs within 2 weeks of the study, cardiovascular disease, cancer, diabetes, average alcohol consumption >2 units a day (1 unit = 12 g), hypersensitivity to any of the tested ingredients, or any other condition that the physician principal investigator (PI) believed could put the subject at risk.

After giving written informed consent, subjects provided a blood sample for a complete metabolic panel (blood chemistry and hematology), which the PI reviewed before the start of the study. On the test day, each subject's medical history was taken, and a brief medical exam was performed. Subjects were then randomized to one of three test groups to receive either a single dose of ibuprofen 400 mg; a single dose of hops resin 450 mg; or four doses of hops powder 300 mg, one dose every 2 hours for 6 hours (Table 4.2).

Table 4.2   Treatment and Control Groups

Group 1 (n = 8): ibuprofen 400 mg

Group 2 (n = 5): hops resin 450 mg

Group 3 (n = 6): hops powder 300 mg q2h X 4

Test Products

Hops Prototypes

The flowering tops and cones of hops (*Humulus lupulus* L., Cannabinaceae) have long been used as ingredients in the beer brewing process to impart bitterness and aroma to the beverage, and to act as a preservative. These effects are due to the presence in hops of a volatile oil, composed chiefly of "alpha" and "beta acids," also known as humulone and lupulone, respectively. These compounds are flavonoids with a prenylated chalcone structure.[48] Humulone has been shown to inhibit TNF-alpha-induced Cox-2 gene induction in a mouse osteoblastic cell model with an IC50 of 30 nM, which suggests that a high-alpha acid hops may possess Cox-inhibiting properties. Hops used for beer making contains typically 5 to 18% alpha acids; the concentrated extract used in the clinical study contained 80% alpha acids. Two different forms of the hops prototype were developed for this study, a resin form packaged in a gelcap and a powder form packaged in a two-piece hardshell. A single tablet of either formulation contained 150 mg alpha acids from hops extract oleoresin, as well as 0.5 mg astaxanthin, a marine antioxidant produced by microalgae, which has been found to limit exercise-induced skeletal muscle damage in mice and possess potent antioxidant effects in humans.[49] The resin formulation also contained the liquid-phase phenolic dieterpene carnosic acid, and the powder formulation contained the solid-phase cafeoyl derivative rosmarinic acid, both antioxidants found in rosemary.[50] (Table 4.3). Both products were test formulas manufactured by the Nutrilite Division of Access Business Group LLC using IsoOxygene™ hops extract provided by Lipoprotein Technologies, Inc. (Bodega, CA).

Subjects provided either 7 (hops resin 450 mg and ibuprofen group) or 9 (hops powder 1200 mg in 4 divided doses group) blood draws and took either 1 or 4 doses of product, over a 9-h test-day.

Ibuprofen

Ibuprofen is a commonly used, non-prescription NSAID with fever-reducing and pain-relieving properties; reduced Cox-2 enzyme activity is largely responsible for its anti-inflammatory and pain-relieving effects. In this study an OTC commercial product (Advil™) was used; a dose of two tablets (delivering 400 mg ibuprofen) was used as positive control.

Table 4.3 Active and Carrier Ingredients in the Two Formulations of Hops Extract Tablets

Product	Active ingredients, per tablet	Carrier
Hops powder	• 150 mg hops alpha acids (from hops [*humulus lupulus*, leaves, flowers] extract) • 10 IU natural Vitamin E (from 8 mg d-alpha tocopherol acetate) • 0.5 mg astaxanthin (from *Haematococcus pluvialis* algal extract) • 1.8 mg rosmarinic acid (from rosemary [*Rosemarinus officinalis*, leaves, flowers] extract)	• <50% maltodextrin, • 10% calcium silicate • 6% vegetable stearine
Hops resin	• 150 mg hops alpha acids (from hops [*humulus lupulus*, leaves, flowers] extract) • 10 IU natural Vitamin E (from 8 mg d-alpha tocopherol acetate) • 0.5 mg astaxanthin (from *Haematococcus pluvialis* algal extract) • 1.5 mg carnosic acid (from rosemary [*Rosemarinus officinalis*, leaves, flowers] extract)	• 100 mg olive oil

STATISTICAL METHODS

In vivo results are shown as means (plus and minus standard error of the mean) of duplicate treatments assayed in duplicate. *Ex vivo* results are expressed as group means with standard errors. The degree of *ex vivo* Cox-1 and Cox-2 inhibition caused by plasma samples was calculated as a percentage of the activity measured at baseline (before any product consumption). The areas over the curve (AOC), representing the total Cox inhibition over 9 hours, was calculated with Prism (Graphpad Software, Inc., Version 4 for Windows). Larger AOCs correspond to greater levels of Cox inhibition. Cox-2 selectivity was calculated by dividing Cox-1 AOC by Cox-2 AOC. Data were tested for normality using the Kolmogorov-Smirnov test with the Lilliefors method for P value approximation. Cox-2 selectivity (Cox-1/Cox-2 ratio) within groups was compared to a hypothetical value of 1 (representing equipotent inhibition of either enzyme) with a Wilcoxon Signed Rank Test. A two-tailed P value <0.05 was considered significant. Normally distributed Cox-2 inhibitory potency (Cox-2 AOC) and selectivity (Cox-1 AUC/Cox-2 AOC) were analyzed by individual subjects first and then averaged by groups and compared with a one-way analysis of variance, followed by Dunn's Multiple Comparisons Test of the four treatment groups to the ibuprofen control group if the overall P value was <0.05. Selectivity comparisons between products were calculated with a one-way anova and Bonferroni's Multiple Comparisons Test if the overall P value was <0.05. Cox-2 inhibitory potency and selectivity data were analyzed for outliers beforehand with Grubb's test. A single outlier identified in this fashion was removed from calculations.

RESULTS

In Vitro Cox Inhibition

Botanicals were selected for *in vitro* testing based on one or more of the following criteria: known Cox-2 inhibitor; known anti-inflammatory action; or traditional for pain relief. *In vitro* results are presented, separated by degree of Cox-2 inhibitory activity for clarity, in Panels A (IC_{80} > 50 mcg/ml), B (IC_{80} = 10-50 mcg/ml), and C (IC_{80} <10 mcg/ml) of Figure 4.1.

A product was deemed unsuitable for further development if a concentration greater than 50 mcg/ml was required to achieve 80% Cox-2 inhibition. This 50 mcg/ml *in vitro* concentration roughly translates to 500 mg per the 5 L of a human whole body blood volume; a crude guideline akin to the 1 g/kg guideline recommended for hippocratic screening of botanical products in rats.[17]

Consistent with known activity against Cox enzymes, aspirin, ibuprofen, and celecoxib (Panel C) were markedly effective in inhibiting PGE_2 secretion in response to PMA stimulation, with celecoxib being the most effective with an IC_{80} of 0.5–0.9 mcg/ml (see Table 4.1 for IC_{50} and IC_{80} summary). Of the test products, turmeric root, grains of paradise (*A. stipulatum*), wild blueberry extract, and the hops extract were the most effective in inhibiting pro-inflammatory cytokine PGE_2 with presumed IC_{80} values <10 mcg/ml.

Ex Vivo Cox Inhibition

Validation Study

Results from *ex vivo* assays performed at ABG and at William Harvey Research Limited were compared with a 2-tailed Pearson correlation test; analysis was performed twice, once for each Cox isoenzyme, using data pooled from both NSAIDs (ibuprofen and celecoxib) together. For Cox-1, R squared was 0.67 (p <0.01) and for Cox-2, R squared was 0.92 (p <0.001).

Clinical Study

Cox-2 Inhibitory Potency

The best performing agent from the *in vitro* assay was then evaluated in an *ex vivo* assay. The inhibitory potency of the test products at any given time within the 9-h sampling period is shown in Figure 4.2; Cox inhibition was achieved within 1 to 2 hours of dosing with the 2 hops dosage forms (450 mg resin in a single dose, and 1200 mg powder in 4 divided doses). There were no significant differences in total Cox-2 inhibitory potency over 9 hours (P = 0.88) (Figure 4.3).

Total Cox-1 and Cox-2 inhibition over the 9-h sampling period are shown as integrated areas over the inhibition curve (AOC) in Table 4.4, along with Cox-2 selectivity ratios, calculated by dividing the Cox-1 by the Cox-2 AOC.

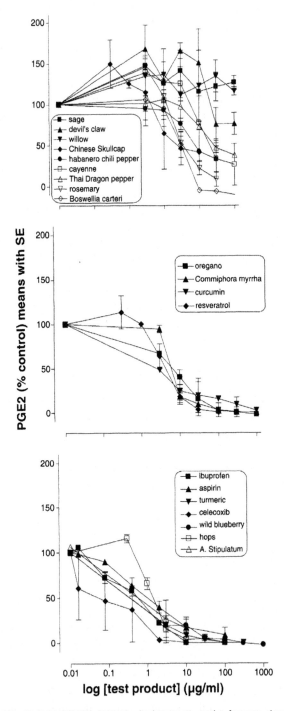

Figure 4.1   *In vitro* Cox-2 inhibition by botanical extracts and reference drugs.

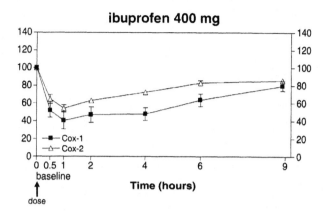

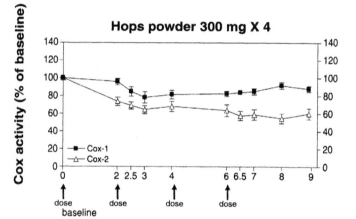

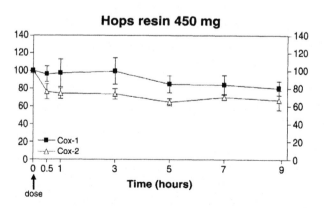

Figure 4.2 Cox-1 and Cox-2 inhibitory activity over 9 hours as a percent of baseline activity.

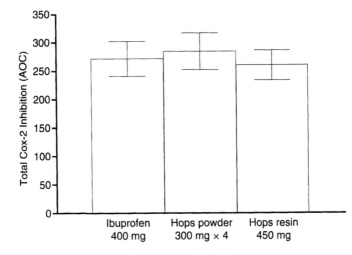

Figure 4.3   Total Cox-2 inhibitory potency (percent inhibition × hours) over 9 hours.

Table 4.4   Area over the Inhibition Curve Values Obtained from the Curves Shown in Figure 4.2

Treatment	AOC			Statistics Cox Selectivity		
	Cox-1	Cox-2	Ratio	Within products	Compared to control	Between products
Hops powder 300 mg × 4	113.5 ± 18.2	284.5 ± 32.2	0.44 ± 0.09	*	*	a
Hops resin 450 mg	129.8 ± 70.3	260.2 ± 26.2	0.42 ± 0.18	*	*	a
Ibuprofen 400 mg	383.0 ± 69.9	271.7 ± 30.6	1.50 ± 0.31	*	(control)	b

* = $P < 0.05$. Selectivity comparisons between products: One-Way Anova $P = 0.01$; ratios not sharing a letter are significantly different.

Cox-2 Selectivity

Consistent with *in vitro* observations, plasma from subjects who had taken ibuprofen 400 mg exhibited significant Cox-1 selectivity with a ratio of 1.50 over 9 hours, a result comparable to the reported *ex vivo* selectivity ratio of 1.58 over 8 hours for the unselective Cox inhibitor Naproxen 500 mg.[34]

Plasma from subjects who had taken either a single dose of hops resin 450 mg, or hops powder 1200 mg in 4 divided doses, showed statistically significant Cox-2 selectivity, which was in turn significantly different from the Cox-1 selectivity of the ibuprofen reference group. The 9-hour Cox-2 selectivity ratios of hops powder 1200 mg and hops resin 450 mg (0.44 and 0.42, respectively) are in the range of

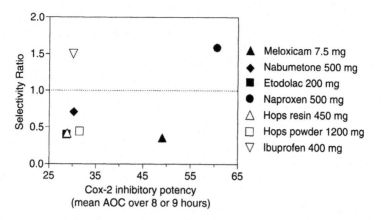

Figure 4.4   Cox-2 selectivity and inhibitory potency of test products and reference drugs.

the 8-hour Cox-2 selectivity ratios reported for plasma taken from subjects who had taken the Cox-2 selective drugs etodolac 200 mg, meloxicam 7.5 mg, and nabumetone 500 mg (0.40, 0.36, and 0.71, respectively).[34]

Since Cox-2 inhibitory selectivity and potency are both important features of Cox-inhibition, it would be of interest to compare on that basis the test products with known pharmaceutical Cox-2 inhibitors. In Figure 4.4, the Cox-2 selectivity ratios and Cox-2 Areas Over the inhibition Curve obtained from various Cox-2 inhibitors reported by Giuliani and colleagues are compared with those from the present study. (The AOCs are represented as averages per hour to make a comparison between the 8-hour and 9-h sampling periods possible.) Thus, in Figure 4.4, plasma from subjects who had taken ibuprofen 400 mg inhibited Cox-2 by a mean of about 30% over 9 hours, with a selectivity ratio between 1 and 2; plasma from subjects who had taken naproxen was much more potent in inhibiting Cox-2 over 8 hours, but it was not more Cox-2 selective; and meloxicam proved less potent but more selective than naproxen. Plasma samples from subjects who had taken hops resin 450 mg exhibited a mean Cox-2 inhibition and selectivity that overlaps with the corresponding value from subjects who had taken etodolac 200 mg in the Giuliani et al. study.[34] The results from the present study, considered in conjunction with published data using the same assay method, supports the pharmacological effectiveness of hops resin 450 mg, or hops powder 1200 mg in 4 divided doses, to inhibit Cox-2 with a potency and a selectivity comparable to that of a known Cox-2 selective drug.

DISCUSSION

In vitro Cox-2 inhibitory potency and selectivity data often fail to predict clinical usefulness. In order to predict *in vivo* performance, they must be related to bioavailability data. For example, although the wild blueberry and turmeric test products performed well *in vitro*, bioavailability data from the literature suggests that the presumed active ingredients in such products, anthocyanins and curcuminoids, respectively, would possess only local anti-inflammatory activity (i.e., in the GI tract) and would not provide systemic Cox-2 inhibition. While hops has previously been found to exert Cox-2 inhibition at the level of gene expression,[29,30] this study represents the first time a hops extract consumed in the form of a dietary supplement has been shown to exert *ex vivo* Cox-2 inhibition in humans with a potency, summed over 9 hours, comparable to that obtained with a dose of ibuprofen 400 mg, but with a theoretically more favorable selectivity, comparable to known Cox-2 selective drugs. The possible clinical utility of hops on this basis — that is, Cox inhibition with a predilection toward Cox-2 rather than Cox-1 — remains to be determined. The term "specific Cox-2 inhibitor" should be reserved for those agents with demonstrated lower GI toxicity in patients.[51] While the roughly three-fold difference in Cox-2 selectivity between the dietary supplements and ibuprofen was modest, this small difference could yet yield clinical benefit. While the peak Cox-2 inhibition achieved by one of the test products (hops powder 1200 mg in 4 divided doses) was equivalent to that achieved by ibuprofen 400 mg, the time course of this maximal effect was different: hops caused a maximum 45% inhibition of Cox-2 at 8 hours following four doses over 6 hours, whereas ibuprofen caused the same degree of inhibition within 1 hour of dosing. Clinical testing would be required to establish clinical pain-relieving efficacy and to test for possible pain relief differences between the dietary supplement prototype and ibuprofen based on differences in time to maximum pharmacological effect.

A safety study involving more subjects, and a risk–benefit analysis based on actual use in different populations, would be required to assess the potential use of a hops-based dietary supplement in a clinical setting (e.g., to treat benign pain) or as a long-term supplement to maintain general health and reduce the risk of cancer.

Although botanical anti-inflammatory dietary supplements have generally not been found to be as potent as analgesic drugs, it has been suggested that they could at least be used to decrease the consumption of NSAIDs, with a concomitant decrease in healthcare expense and risk of gastric side effects.[14] Another possible application is use by healthy persons who engage in strenuous exercise, which increases oxidative stress and microscopic tissue damage,[52] and gives rise to a non-disease yet painful state that could potentially respond to an anti-inflammatory product. It is clear, however, that if a Cox-2 selective dietary supplement could be produced and marketed at a lower cost to the consumer, more could benefit from the improved safety profile of Cox-2 selective over non-selective Cox-inhibiting agents.

In summary, the present study provides enough evidence to support the utility of a powdered hops extract standardized to alpha acids as an anti-inflammatory dietary supplement with a predilection toward inhibiting the Cox-2 over the Cox-1 isoform of the cyclooxygenase enzyme. Whether this predilection yields an improved GI safety profile, and whether the mechanism of action of hops would be more suited to local (GI tract) anti-inflammatory and anti-cancer effects such as with turmeric curcuminoids, or to pain relief as with willow bark extracts, remains to be determined.

REFERENCES

1. Doyle, G., Furey, S., Berlin, R., Cooper, S., Jayawardena, S., Ashraf, E., et al., Gastrointestinal safety and tolerance of ibuprofen at maximum over-the-counter dose, *Aliment Pharmacol Ther,* 1999;13(7):897–906.
2. Erlacher, L., Wyatt, J., Pflugbeil, S., Koller, M., Ullrich, R., Vogelsang, H., et al., Sucrose permeability as a marker for NSAID-induced gastroduodenal injury, *Clin Exp Rheumatol,*1998;16(1):69–71.
3. Brooks, P., Emery, P., Evans, J.F., Fenner, H., Hawkey, C.J., Patrono, C., Smolen, J., Breedveld, F., Day, R., Dougados, M., Ehrich, E.W., Gijon-Banos, J., Kvien, T.K., Van Rijswijk, M.H., Warner, T., and Zeidler, H., Interpreting the clinical significance of the differential inhibition of cyclooxygenase-1 and cyclooxygenase-2, *Rheumatology (Oxford),* 1999 Aug;38(8):779–788.
4. Mamdani, M., Rochon, P.A., Juurlink, D.N., Kopp, A., Anderson, G.M., Naglie, G., Austin, P.C., and Laupacis, A., Observational study of upper gastrointestinal haemorrhage in elderly patients given selective cyclo-oxygenase-2 inhibitors or conventional non-steroidal anti-inflammatory drugs, *BMJ,* 2002 Sep 21;325(7365):624.
5. Watson, D.J., Yu, Q., Bolognese, J.A., Reicin, A.S., and Simon, T.J., The upper gastrointestinal safety of rofecoxib vs. NSAIDs: an updated combined analysis, *Curr Med Res Opin,* 2004 Oct;20(10):1539–1548.
6. Garner, S., Fidan, D., Frankish, R., Judd, M., Shea, B., Towheed, T., Wells, G., and Tugwell, P., Celecoxib for rheumatoid arthritis, *Cochrane Database Syst Rev,* 2002;(4):CD003831.
7. Maetzel, A., Krahn, M., and Naglie, G., The cost effectiveness of rofecoxib and celecoxib in patients with osteoarthritis or rheumatoid arthritis, *Arthritis Rheum,* 2003 Jun 15;49(3):283–292.
8. Warner, T.D., Giuliano, F., Vojnovic, I., Bukasa, A., Mitchell, J.A., and Vane, J.R., Nonsteroid drug selectivities for cyclo-oxygenase-1 rather than cyclo-oxygenase-2 are associated with human gastrointestinal toxicity: a full in vitro analysis, *Proc Natl Acad Sci USA,* 1999 Jun 22;96(13):7563–7568.
9. Mamdani, M., Juurlink, D.N., Lee, D.S., Rochon, P.A., Kopp, A., Naglie, G., Austin, P.C., Laupacis, A., and Stukel, T.A., Cyclo-oxygenase-2 inhibitors versus non-selective non-steroidal anti-inflammatory drugs and congestive heart failure outcomes in elderly patients: a population-based cohort study, Lancet 2004 May 29;363(9423):1751–1756.
10. Garner, S., Fidan, D., Frankish, R., and Maxwell, L., Rofecoxib for osteoarthritis, *Cochrane Database Syst Rev,* 2005 Jan 25;1:CD005115.
11. Rampal, P., Moore, N., Van Ganse, E., Le Parc, J.M., Wall, R., Schneid, H., and Verriere, F., Gastrointestinal tolerability of ibuprofen compared with paracetamol and aspirin at over-the-counter doses, *J Int Med Res,* 2002 May–Jun;30(3):301–308.
12. Warner, T.D. and Mitchell, J.A., Cyclooxygenases: new forms, new inhibitors, and lessons from the clinic, *FASEB J,* 2004 May;18(7):790–804.
13. Chrubasik, S., Eisenberg, E., Balan, E., Weinberger, T., Luzzati, R., and Conradt, C., Treatment of low back pain exacerbations with willow bark extract: a randomized double-blind study, *Am J Med,* 2000;109(1):9–14.
14. Chrubasik, S., Kunzel, O., Black, A., Conradt, C., and Kerschbaumer, F., Potential economic impact of using a proprietary willow bark extract in outpatient treatment of low back pain: an open non-randomized study, *Phytomedicine,* 2001;8(4):241–251.
15. Schmid, B., Kotter, I., and Heide, L., Pharmacokinetics of salicin after oral administration of a standardised willow bark extract, *Eur J Clin Pharmacol,* 2001 Aug;57(5):387–391.

16. Morgan, G., Should aspirin be used to counteract "salicylate deficiency"? *Pharmacol Toxicol,* 2003 Oct;93(4):153–155.
17. Malone, M.H., The pharmacological evaluation of natural products — general and specific approaches to screening ethnopharmaceuticals, *J Ethnopharmacol,* 1983 Aug;8(2):127–147.
18. Kornman, K.S., Martha, P.M., and Duff, G.W., Genetic variations and inflammation: a practical nutrigenomics opportunity, *Nutrition,* 2004 Jan;20(1):44–49.
19. Mitchell, J.A. and Warner, T.D., Cyclo-oxygenase-2: pharmacology, physiology, biochemistry and relevance to NSAID therapy, *Br J Pharmacol,* 1999;128(6):1121–1132.
20. Barden, J., Edwards, J.E., McQuay, H.J., Wiffen, P.J., and Moore, R.A., Relative efficacy of oral analgesics after third molar extraction, *Br Dent J,* 2004 Oct 9;197(7):407–411; discussion 397.
21. Xu, X.M., Sansores-Garcia, L., Chen, X.M., Matijevic-Aleksic, N., Du, M., and Wu, K.K., Suppression of inducible cyclooxygenase 2 gene transcription by aspirin and sodium salicylate, *Proc Natl Acad Sci USA,* 1999 Apr 27;96(9):5292–5297.
22. Wu, K.K., Aspirin and salicylate: an old remedy with a new twist, *Circulation,* 2000 Oct 24;102(17):2022–2023.
23. Dixon, D.A., Regulation of COX-2 expression in human cancers, *Prog Exp Tumor Res,* 2003;37:52–71.
24. Ristimaki, A., Cyclooxygenase 2: from inflammation to carcinogenesis, *Novartis Found Symp,* 2004;256:215–221; discussion 221–226, 259–269.
25. Davies, G.L., Cyclooxygenase-2 and chemoprevention of breast cancer, *J Steroid Biochem Mol Biol,* 2003 Sep;86(3–5):495–499.
26. Koistinaho, J. and Chan, P.H., Spreading depression-induced cyclooxygenase-2 expression in the cortex, *Neurochem Res,* 2000 May;25(5):645–651.
27. Andreasson, K.I., Savonenko, A., Vidensky, S., Goellner, J.J., Zhang, Y., Shaffer, A., Kaufmann, W.E., Worley, P.F., Isakson, P., and Markowska, A.L., Age-dependent cognitive deficits and neuronal apoptosis in cyclooxygenase-2 transgenic mice, *J Neurosci,* 2001 Oct 15;21(20):8198–8209.
28. Lu, J., Ho, C.T., Ghai, G., and Chen, K.Y., Differential effects of theaflavin monogallates on cell growth, apoptosis, and Cox-2 gene expression in cancerous versus normal cells, *Cancer Res,* 2000 Nov 15;60(22):6465–6471.
29. Yamamoto, K., Wang, J., Yamamoto, S., and Tobe, H., Suppression of cyclooxygenase-2 gene transcription by humulon of beer hop extract studied with reference to glucocorticoid, *FEBS Lett,* 2000 Jan 14;465(2–3):103–106.
30. Yamamoto, K., Wang, J., Yamamoto, S., and Tobe, H., Suppression of cyclooxygenase-2 gene transcription by humulon, *Adv Exp Med Biol,* 2002;507:73–77.
31. Calixto, J.B., Beirith, A., Ferreira, J., Santos, A.R., Filho, V.C., and Yunes, R.A., Naturally occurring antinociceptive substances from plants, *Phytother Res,* 2000 Sep;14(6):401–418.
32. Sampson, J.H., Phillipson, J.D., Bowery, N.G., O'Neill, M.J., Houston, J.G., and Lewis, J.A., Ethnomedicinally selected plants as sources of potential analgesic compounds: indication of in vitro biological activity in receptor binding assays, *Phytother Res,* 2000 Feb;14(1):24–29.
33. Giuliano, F. and Warner, T.D., Ex vivo assay to determine the cyclooxygenase selectivity of non-steroidal anti-inflammatory drugs, *Br J Pharmacol,* 1999; 126(8):1824–1830.
34. Giuliano, F., Ferraz J.G., Pereira R., de Nucci G., and Warner T.D., Cyclooxygenase selectivity of non-steroid anti-inflammatory drugs in humans: ex vivo evaluation, *Eur J Pharmacol,* 2001;426(1–2):95–103.

35. Miller, M.J., Vergnolle, N., McKnight, W., Musah, R.A., Davison, C.A., Trentacosti, A.M., Thompson, J.H., Sandoval, M., and Wallace, J.L., Inhibition of neurogenic inflammation by the Amazonian herbal medicine sangre de grado, *J Invest Dermatol,* 2001 Sep;117(3):725–730.

36. Singh, S., Majumdar, D.K., and Rehan, H.M., Evaluation of anti-inflammatory potential of fixed oil of Ocimum sanctum (Holybasil) and its possible mechanism of action, *J Ethnopharmacol,*1996;54(1):19–26.

37. Godhwani, S., Godhwani, J.L., and Vyas, D.S., Ocimum sanctum: an experimental study evaluating its anti-inflammatory, analgesic and antipyretic activity in animals, *J Ethnopharmacol,* 1987;21(2):153–163.

38. Plummer, S.M., Hill, K.A., Festing, M.F., Steward, W.P., Gescher, A.J., and Sharma, R.A., Clinical development of leukocyte cyclooxygenase 2 activity as a systemic biomarker for cancer chemopreventive agents, *Cancer Epidemiol Biomarkers Prev,* 2001;10(12):1295–1299.

39. Ireson, C., Orr, S., Jones, D.J., Verschoyle, R., Lim, C.K., Luo, J.L., et al., Characterization of metabolites of the chemopreventive agent curcumin in human and rat hepatocytes and in the rat in vivo, and evaluation of their ability to inhibit phorbol ester-induced prostaglandin E2 production, *Cancer Res,* 2001;61(3):1058–1064.

40. Plummer, S.M., Holloway, K.A., Manson, M.M., Munks, R.J., Kaptein, A., Farrow, S. et al., Inhibition of cyclo-oxygenase 2 expression in colon cells by the chemopreventive agent curcumin involves inhibition of NF-kappaB activation via the NIK/IKK signalling complex, *Oncogene,* 1999;18(44):6013–6020.

41. Rao, C.V., Rivenson, A., Simi, B., and Reddy, B.S., Chemoprevention of colon carcinogenesis by dietary curcumin, a naturally occurring plant phenolic compound, *Cancer Res,* 1995;55(2):259–266.

42. Huang, M.T., Lysz, T., Ferraro, T., Abidi, T.F., Laskin, J.D., and Conney, A.H., Inhibitory effects of curcumin on in vitro lipoxygenase and cyclooxygenase activities in mouse epidermis, *Cancer Res,* 1991;51(3):813–819.

43. Murakami, A., Takahashi, D., Kinoshita, T., Koshimizu, K., Kim, H.W., Yoshihiro, A., et al., Zerumbone, a Southeast Asian ginger sesquiterpene, markedly suppresses free radical generation, proinflammatory protein production, and cancer cell proliferation accompanied by apoptosis: the alpha,beta-unsaturated carbonyl group is a prerequisite, *Carcinogenesis,* 2002;23(5):795–802.

44. Tjendraputra, E., Tran, V.H., Liu-Brennan, D., Roufogalis, B.D., and Duke, C.C., Effect of ginger constituents and synthetic analogues on cyclooxygenase-2 enzyme in intact cells, *Bioorg Chem,* 2001;29(3):156–163.

45. Cuendet, M. and Pezzuto, J.M., The role of cyclooxygenase and lipoxygenase in cancer chemoprevention, *Drug Metabol Drug Interact,* 2000;17(1–4):109–157.

46. Altman, R.D. and Marcussen, K.C., Effects of a ginger extract on knee pain in patients with osteoarthritis, *Arthritis Rheum,* 2001 Nov;44(11):2531–2538.

47. Mitchell, J.A., Belvisi, M.G., Akarasereenont, P., Robbins, R.A., Kwon, O.J., Croxtall, J., Barnes, P.J., and Vane, J.R., Induction of cyclo-oxygenase-2 by cytokines in human pulmonary epithelial cells: regulation by dexamethasone, *Br J Pharmacol,* 1994 Nov;113(3):1008–1014.

48. Yilmazer, M., Stevens, J.F., Deinzer, M.L., and Buhler, D.R., In vitro biotransformation of xanthohumol, a flavonoid from hops (Humulus lupulus), by rat liver microsomes, *Drug Metab Dispos,* 2001 Mar;29(3):223–231.

49. Guerin, M., Huntley, M.E., and Olaizola, M., Haematococcus astaxanthin: applications for human health and nutrition, *Trends Biotechnol,* 2003 May;21(5):210–216.

50. del Bano, M.J., Lorente, J., Castillo, J., Benavente-Garcia, O., del Rio, J.A., Ortuno, A., Quirin, K.W., and Gerard, D., Phenolic diterpenes, flavones, and rosmarinic acid distribution during the development of leaves, flowers, stems, and roots of Rosmarinus officinalis. Antioxidant activity, *J Agric Food Chem,* 2003 Jul 16;51(15):4247–4253.
51. Vane, J.R. and Warner, T.D., Nomenclature for COX-2 inhibitors, *Lancet,* 2000 Oct 21;356(9239):1373–1374.
52. Niess, A.M., Dickhuth, H.H., Northoff, H., and Fehrenbach, E., Free radicals and oxidative stress in exercise — immunological aspects, *Exerc Immunol Rev,* 1999;5:22–56.

CHAPTER 5

Lipid Peroxidation, Gene Expression, and Resveratrol: Implications in Atherosclerosis

Ozgur Kutuk, Dilek Telci, and Huveyda Basaga

CONTENTS

INTRODUCTION

Cellular signaling provides an organizational component to the multitude of biochemical reactions that occur continuously for individual cells for the maintenance of life. These logistic signals travel between cells via the blood stream and locally inside the tissues. On the membrane, signals are transduced and travel into the interior where they become integrated with other signals and finally target specific processes controlling cell division, differentiation, metabolism, and other cellular functions.

Oxidatively modified low-density lipoproteins and end products of lipid perox-idation have all been shown to affect cellular processes by modulation of signal transduction pathways, hence effecting the nuclear transcription of genes. Oxidized low-density lipoproteins (oxLDL), oxy-sterols, and 4-hydroxynonenal (4-HNE) have been widely investigated and these species are thought to contribute to atherogenesis and development of atherosclerotic plaque. Diets rich in phytochemicals have been demonstrated to modulate intracellular signaling pathways invoked by oxidized lipids and may thus provide a molecular basis for prevention. Among those studied, resveratrol, a polyphenolic compound found in red wine, has been extensively studied for its antioxidant and preventive properties in atherosclerosis.

This review aims to provide an update on the molecular mechanisms underlying resveratrol activity, with special focus on its effect on signaling cascades mediated by oxidized lipids and their breakdown products.

LIPID PEROXIDATION AND CELL SIGNALING

Cellular signaling is mediated by molecules expressed on the cell surface, in the case of cell-cell interactions, or secreted directly to the extracellular environment. Most of these stimuli consist of chemical compounds that transport biologic infor-mation, the "first messengers" or ligands. They are synthesized and released by signaling cells, and produce a specific response in target cells that have a receptor for that particular signal. These signaling molecules involve proteins, amino acids, steroids, fatty acids, and gases such as nitric oxide. Most of the signaling molecules are hydrophilic and therefore cannot pass through the plasma membrane directly. Instead they bind to receptors on target cell membranes.

All cells receive and process signals not only from the plasma membrane but also from different compartments within the cells. These signals involve alterations in the availability of nutrients, growth factors, cytokines, as well as oxidative, heat, or mechanical stresses. Specific pathways receive signals simultaneously and trans-mit them to their corresponding targets; this refined type of cellular information transfer is not through linear freestanding pathways but through strictly controlled signaling networks.[1,2] The dynamic characteristic of cells enables them to interpret and respond to environmental or intracellular changes, a phenomenon that is fun-damental to life (Figure 5.1).

Receptors are the gatekeepers for cellular signaling, and they exhibit specific and high-affinity binding to their corresponding ligands. Once activated by an extracellular ligand, these transmembrane proteins on the cell surface initiate a series of signaling events that result in various cellular responses depending on the nature of the stimuli. However some receptors are located inside the cells, and they are targets for small, hydrophobic first messengers, which are able to diffuse across the plasma membrane.[3]

In most cases, steroid hormones (estradiol, testosterone) and other hydrophobic small signaling molecules act via endocrine signaling cascade. They are synthesized and secreted into the circulation by specialized endocrine cells, and they bind to their corresponding intracellular receptors following direct diffusion across the

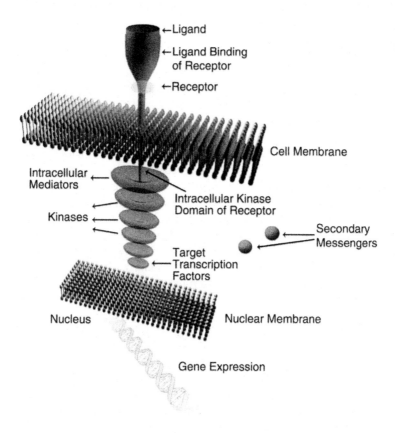

Figure 5.1 Signal transduction mechanism.

plasma membrane. These intracellular receptors may be both in the cytosol or nucleus and, upon stimulation by the signaling hormone, they bind to unique hormone response elements in the promoter region of specific target genes. In addition to this classical hormone-signaling model, recent studies implicate ligand-independent and non-nuclear surface receptor-dependent models for hormone receptors. Although these non-traditional actions of hormone receptors need further mechanistic investigation, these pathways seem to be involved in many physiological and pathological cellular responses.[4-6]

The response to a molecular signal is mainly determined both by the properties of the ligand and receptor as described above, but interestingly, different cell types can respond differently to the same stimuli. The main reason for the divergence of cellular responses to the same ligand-receptor modules can be explained by the complexity of a third factor involved in cellular signaling; intracellular signaling networks involving co-activators and repressors.[2,7]

Briefly, cellular signaling pathways act on multiple intracellular targets, ranging from gene expression to protein localization. The complexity of cellular signaling arises from the large number of gene products involved, simultaneous activation of distinct signaling cascades, coexistence of different protein-protein interactions, relative abundance, as well as intracellular localization of adaptor proteins and duration/amplification of the signal. All these factors, in the end, determine the nature of the response to the signal; proliferation, differentiation, gene expression, cellular migration, cell cycle progression/arrest or apoptosis.

Signaling by Oxidized Lipids

The regulation of cell function by extrinsic factors occurs through a series of second-messenger signals that are initiated by ligand-receptor interaction and are responsible from the modulation of both the intensity and the nature of cellular responses. In this manner, second messengers exert control over principal cellular events, such as metabolism, secretion, structure, and replication.[7,8]

LDL particles are examples of extrinsic factors that have been shown to trigger a variety of biochemical and cellular events associated with the atherosclerotic process, especially when their lipid moieties are oxidized.[9-11] For example, oxLDLs launch a cascade of intracellular signals through the cAMP (A-kinase) or the inositol phosphate pathway, increasing the cytosolic calcium level of cultured endothelial (EC) and smooth muscle cells (SMC).[12,13] Apart from the effect on membrane-bound enzyme activities, oxLDLs also stimulate other elements of the protein kinase C signaling pathway, inducing tyrosine phosphorylation and activation of epithelial growth factor receptor (EGFR).[14,15]

Modulation of the expression and function of vascular genes is yet another effect exerted by oxLDLs by which these genes may participate in inflammatory and atherosclerotic processes. OxLDL increases the VCAM-1 induction response to a cytokine inflammatory signal in endothelial cells from different vascular beds.[16] More recently, oxLDLs have been shown to stimulate intercellular adhesion molecule (ICAM-1) expression in EC and monocyte adhesion, probably through activation of a specific protein tyrosine kinase (PTK).[17,18]

The activation of nuclear transcription factors has been shown to be necessary for the mechanism by which oxLDLs increase the expression of genes involved in the cell growth, differentiation, and proliferative responses observed in atherosclerosis. In fact, it has been demonstrated that oxLDLs stimulate activator protein-1 (AP-1) DNA–binding activity in fibroblasts, SMCs, and ECs, possibly via JNKs and ERKs, and that lipid peroxidation products participates in this phenomenon.[19] The stimulation of DNA-binding activity of another redox-sensitive transcription factor, nuclear factor κ B (NF-κB), by oxLDL has also been described.[20]

This finding has been confirmed by Tontonoz et al. and Nagy et al., who demonstrated that activation of PPAR by oxLDL is dependent on oxLDL internalization via scavenger receptors. In particular, the activation of the PPAR:RXR heterodimer in myelomonocytic cell lines induces transcriptional upregulation of the scavenger receptor CD36, which in turn promotes uptake of oxLDL. Moreover, oxidative modifications of the lipid moiety of LDL particles lead to the production of derivatives (9-HODE, 13-HODE), which are efficient activators of PPAR. Thus, signaling through a surface receptor may be directly coupled to signaling through a nuclear receptor.[21,22]

The biological responses triggered by oxLDLs are associated with lipid peroxidation derivatives, which are able to induce various pathogenic intracellular signals leading to cellular dysfunction. They interfere with various signaling pathways involving calcium, trimeric G-proteins and cAMP, phospholipase C and D, protein kinase C, and MAP kinase cascade. In summary, oxLDLs exhibit many biochemical effects that are potentially involved in atherogenesis, lipid metabolism, and gene expression of adhesion molecules, of heat shock proteins, cytokines, and growth factors, as well as in cell migration, motility and contractility, cell viability, local immune response, and vasomotor tone.[23,24]

Signaling by Oxidized Steroids

Low-density lipoproteins contain numerous other lipid oxidation products apart from aldehydes that can participate in pathological events leading to atherosclerosis, or to modulation of gene expression. Indeed, oxidized derivatives of the major components of LDL, i.e., cholesterol and cholesterol esters, are of interest in this respect because they are consistently found in human atherosclerotic lesions.[25,26]

Oxysterols account for 3 to 5% of total plasma cholesterol in normocholesterolemic animals, and up to 60 to 70% of total cholesterol in oxLDLs found in hypercholesterolemic animals. Applying oxysterols obtained from oxLDLs to cultivated human promonocytes, (U937) cells upregulates production of the fibrogenic cytokine TGFβ1, pointing to a possible effect of these compounds on gene expression. This possibility is supported by the previous observation that 25-hydroxycholesterol is able to induce the mRNA transcript levels of basic fibroblast growth factor in smooth muscle cells.[27]

Signaling by Aldehydic End Products of Lipid Peroxidation

Among the various end products of the oxidative breakdown of biomembrane polyunsaturated fatty acids, aldehydes have been well characterized for their potential contribution to free radical pathobiology. The group of aldehydes that displays the greatest number of biochemical activities is the 4-hydroxyalkenals. Quantitatively, the most representative hydroxyalkenal in animal tissues is 4-hydroxy-2,3-nonenal (4-HNE). This aldehyde derives from the oxidation of arachidonic acid, as well as linoleic acid.[28] Monoclonal antibodies against 4-HNE protein adducts have positively identified this molecule in human chronic diseases, and several reports pointed to the potential effect of increased levels of 4-HNE on different cell functions.[29–33]

We have previously shown that 4-hydroxynonenal (4-HNE) induced oxidative stress and apoptotic cell death in Swiss 3T3 fibroblasts. Indeed, in vitro treatment of 3T3 fibroblasts with 4-HNE induced a condition of oxidative stress as monitored by the oxidation of dichlorofluorescein diacetate. Further, 4-HNE-treated fibroblasts eventually underwent apoptotic death as determined by differential staining and internucleosomal DNA fragmentation. These observations are consistent with a potential role of lipid peroxidation-derived products in programmed cell death.[34]

It has also shown that enhanced lipid peroxidation or cell treatment with 4-HNE at doses compatible to those detectable *in vivo* induce expression and synthesis of the fibrogenic cytokine transforming growth factor β1 (TGFβ1) in rat liver, in macrophage-derived cell lines.[35,36] In addition, collagen production is induced both in rat liver and in cultivated human stellate cells.[35,37]

Both TGFβ1 and collagen type I genes require an active binding of AP-1 transcription factor for their optimal expression for which both genes (TGFβ1 and collagen type 1) have consensus sequences in their promoter regions.[38,39] The demonstrated action of 4-HNE on the expression of defined genes prompted a reanalysis of the aldehyde's effect on cell signaling.

A marked activation of AP-1 nuclear binding as induced by 1–10 μM 4-HNE, has now been demonstrated in different cytotypes including murine and human macrophages,[40] human hepatic stellate cells,[41] and rat hepatocytes. With regard to another well-studied redox-sensitive transcription factor, NF-κB, 4-HNE does not appear to influence its activation and nuclear translocation in the mentioned cytotypes.[40,41] Moreover, in neuroblastoma-derived PC12 cell line, 4-HNE activates AP-1 and significantly downregulates NF-κB nuclear binding.[42]

Mainly based on the two latter findings, the hypothesis is now put forward that, under defined conditions, the aldehyde acts as an apoptotic trigger, signaling through the MAP kinase pathway (Figure 5.2) and AP-1 binding (Figure 5.3), and simultaneously depressing NF-B activation.

Mitogen-activated protein kinases (MAPKs) and various caspases have been proposed to mediate stress-induced apoptosis in many cell types. In order to get insight into the mechanisms of apoptotic response by 4-HNE in 3T3 fibroblasts, we followed MAPK and caspase activation pathways; 4-HNE induced early activation of JNK1/2 and p38 pathways and ERK1/2.

AP-1 is a transcription factor involved in the regulation of cell growth, differentiation, proliferation, and apoptosis. AP-1 exists as homo- and heterodimers and is activated by growth factors, cytokines, and cellular stress. c-jun and c-fos are two

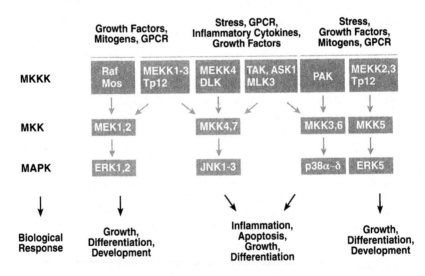

Figure 5.2 MAPK signaling pathways.

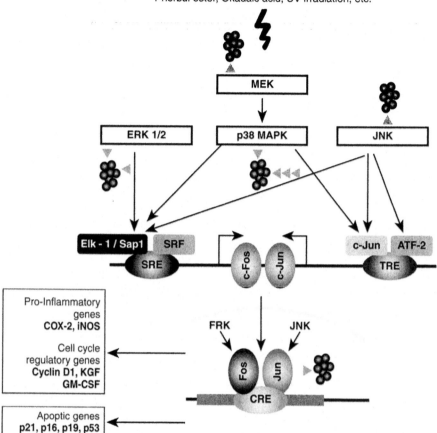

Inflammatory stimuli, Cytokines, Growth factors, ROS, Phorbul ester, Okadaic acid, UV irradiation, etc.

Figure 5.3 Mechanism of AP-1 upregulation and its effect on transcription of a number of genes.

main proteins that form AP-1 transcription factor complex and are targets for phosphorylation by MAP kinases to be transcriptionally active.

Parallel studies from different laboratories on different cytotypes led to the conclusion that the aldehyde strongly activates c-Jun amino-terminal kinases (JNKs), promoting their translocation in the nucleus where JNK-dependent induction of c-jun and AP-1 takes place.[41,43]

It appears likely that in cells other than hepatic stellate cells, 4-HNE activates JNKs through phosphorylation rather than direct binding. Uchida et al., demonstrating a rapid increase in JNK phosphorylation in rat liver epithelial RL34 cells, reached this conclusion.[44] These authors also reported a mild effect of the aldehyde on ERK activity, while Ruef et al. showed a significant activation of ERK1 and ERK2 in 4-HNE treated rat aortic smooth muscle cells.[45] The aldehyde, at concentrations effective on JNKs, also markedly activates novel PKC isoforms, in particular, PKC beta in hepatocytes.[46]

The most striking evidence in support of a pathophysiologic involvement of 4-HNE in gene modulation is the demonstration that the aldehyde externally added could be rapidly recovered within the nuclei of cultured cells. This finding was first obtained in liver stellate cells as reported above,[41] and confirmed in macrophages. This finding paralleled that of increased AP-1 binding.[41]

PHYTOCHEMICALS AND RESVERATROL IN CELL SIGNALING

Structure

Phytonutrients are compounds found in plants, which are not required for normal functioning of the body but which nonetheless have a beneficial effect on health or an active role in the amelioration of diseases. The major families of related phyto-nutrients are flavonoids, phytoestrogens, isothiocyanates, monoterpenes, organosul-fur compounds, saponins, capsaicin, and sterols.[47–49] Many recent molecular and cellular studies have focused on the beneficial effect of dietary phytoestrogens in reducing the risk of cardiovascular diseases and cancer.[49,50] Phytoestrogens are polyphenolic non-streroidal secondary by-products of plant metabolism that play an important role in protecting the plant from environmental stressors and predators. In addition to chemopreventive and antiangiogenic properties, acting as a functional analog of mammalian estrogen 17-b-estradiol (E2), phytoestrogens have potential to gain importance at clinical applications in hormone therapy to prevent menopausal syndromes and osteoporosis.[49,51,52]

Currently, there are four phenolic compounds considered as phytoestrogens: the isoavonoids, lignans, coumestans, and stilbenes. Isoflavonoids are the subclass of flavonoids reported to exist as glucosides. They are hydrolyzed in the gut to aglycones and transported across intestinal epithelial cells.[52,53] The isoavonoids appear in high amounts in soy-based foods.[47,54] Lignans are a diverse class of phenylpropanoid dimers and oligomers that can readily converted to mammalian lignans such as enterodiol and enterolactone by the gut microflora.[55,56] The major source of phytolignans in the diet is from axseed, whole grain breads, vegetables, and tea.[54,57] Coumestans occur predominantly during germination. A large number of coumestans have been identified, few of which show an estrogenic activity. The main compound in this subgroup with estrogenic activity is coumesterol, found mainly in legume sprouts.[58,59]

Stilbenes, like the avonoids, are synthesized via the phenylpropanoid-acetate pathway. Resveratrol (trans-3,5,40-trihydroxystilbene) (Figure 5.4) has been identi-fied as the major active compound of the stilbene phytoestrogens. It exists as trans and cis isomers, however only the trans form has been reported to be estrogenic.[60] The inverse relationship between risk of ischemic heart disease and French-pattern dietary intake, moderate consumption of red wine and consumption of foods high in saturated fats, raised a wave of interest in monitoring the presence of resveratrol in red wine.[61–65] Resveratrol is found only in grape skin, not in grape flesh, making this stilbene the active ingredient of red wine but resulting in low levels in white wine.[66] Resveratrol was also detected in itadori tea and in peanut roots.[67,68]

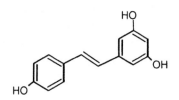

Figure 5.4   Structure of resveratrol.

Pharmacokinetic and Antioxidant Properties

In evaluation of the response of living organisms to resveratrol, it is important to determine the capacity of gastrointestinal absorption and the distribution of resveratrol to various target tissues as well as its metabolism. The bacterial microflora in the ileum and cecum plays an important role in the metabolism and absorption of resveratrol. Kinetics of *trans-* and *cis*-resveratrol (3,4',5-trihydroxystilbene) after oral administration of red wine in rats showed that a fraction of resveratrol (6.5mg/l) was absorbed by rats and can be detected in considerable concentrations in plasma. Peak tissue concentrations recorded at cardiac tissue, liver, and kidneys.[69] A recent study reported albumin as one of the plasmatic carriers of resveratrol. Interaction of resveratrol with albumin is essential for the transporting of the compound in blood circulation and the delivery of resveratrol at the cell surface before cell membrane uptake.[70]

Results from recent studies demonstrated the ability of resveratrol to detoxify reactive oxygen species (ROS). ROS such as hydroxyl radical ($\cdot$ OH), superoxide radical ($O_2^{\cdot-}$), and hydrogen peroxide (H_2O_2) are generated as by-products of molecular cellular events and are involved in initiation of oxidative stress.[71] Cells can tolerate mild oxidative stress and often respond by synthesizing intracellular antioxidants such as glutathione, α-tocopherol, ascorbic acid, β-carotene, and antioxidant enzymes including catalase, superoxide dismutase, and glutathione peroxidase as a part of the defense system against ROS-induced cellular injury. These antioxidants may provide a buffer against oxidative stress by quenching ROS. However, when ROS production outweighs antioxidant defenses, cell injury or even cell death may occur due to oxidative damage to DNA, proteins, and lipids.[72] Resveratrol has been proposed to prevent superoxide-dependent inflammatory response by acting in a similar way to superoxide-dismutase.[73] For example, resveratrol inhibits the detrimental effects of pro-oxidant tert-butyl hydroperoxide in cultured rat embryonic mesencephalic cells.[74] Results from another recent study demonstrate that pre-incubation of macrophages with resveratrol prior to treatments with lipopolysaccharides or phorbol esters exerts a strong inhibitory effect on superoxide radical and hydrogen peroxide production.[75]

The radical scavenging activity of resveratrol has also been shown in a number of *in vitro* studies. Leonard et al. reports that resveratrol is an efficient $\cdot$ OH and O2$\cdot^-$ radical scavenger.[76] By using hexavalent chromium (Cr[IV]), they have shown the ability of resveratrol to scavenge radicals produced by the enzymatic reaction with

Cr[IV]. Consistent results were also obtained when *in vitro* synthesized resveratrol derivatives showed free radical scavenging activity.[50]

Another aspect of free radical damage involves the lipid peroxidation of plasma lipoproteins and cellular membranes, as mentioned in the first part of this review. Oxidized polyunsaturated fatty acids in LDL show chemostatic properties for monocytes and leading to endothelial dysfunction and vascular smooth muscle cell proliferation.[77] The oxidation also affects LDL binding to collagen, and stimulates connective tissue formation and monocyte-endothelial cell interaction.[78–80] Thus, oxidized lipids in LDL may increase the permeability of endothelial cells leading to accumulation and retention of intracellular and extracellular lipids at subendothelial membrane and infiltration of pro-inflammatory monocytes and lymphocytes, which are believed to play a major role in development of atherosclerosis.[81–84] Atherosclerosis is a chronic inflammatory condition characterized by early fatty degeneration and high levels of circulating LDL, monocytes, and lymphocytes, leading to inflammation at the vessel wall.[71]

Given the proposed role of oxLDL in the development of atherosclerosis, studies have been performed to evaluate the role of antioxidants such as resveratrol on inhibition of LDL oxidation. For example, short-term consumption of grape juice by patients with coronary heart disease increases the mean lag time of LDL oxidation *in vitro*.[85–87]

In vitro, resveratrol behaves as a slight antioxidant against oxidation of fatty acids of sunflower oil and rapeseed oils.[88] Using a copper chloride-induced LDL oxidative system, addition of resveratrol prolonged the lag time preceding the onset of LDL oxidation in a dose-dependent manner.[89] These findings were confirmed in other assays showing the ability of resveratrol to reduce the oxidative alterations of lipid and protein moieties of LDL. It lowers the non-specific uptake of oxLDL by macrophages, hence preventing the formation of lipid-laden foam cells.[90]

Resveratrol also inhibits the peroxidation of membrane lipids that destroy the phospholipid bilayer of the cellular membrane, resulting in irreversible cell damage. Resveratrol pretreatment of a rat model of dopaminergic neuronal cells prevents lipid oxidation caused by iron and ethanol.[91] In human primary cell cultures and rat liver microsomes, Stivala et al. showed that resveratrol inhibits the lipid peroxidation induced by free radical donors.[92] Furthermore, lipid peroxidation in cell membrane of macrophage tumor cells caused by exposure to $\cdot$OH radicals is inhibited by resveratrol.[76]

Resveratrol in Cell Signaling

Resveratrol has been reported to protect against H_2O_2 and β-amyloid-induced oxidative cell death in rat pheochromocytoma cells by attenuating intracellular ROS accumulation and restoring the levels of some marker proteins of apoptosis, such as Bax, Bcl-X_L, JNK, and PARP.[93,94] Leonard et al. found that resveratrol treatment of macrophage tumor cells inhibits the DNA damage due to $\cdot$OH radicals.[76] In experimental models of oxidant stress-mediated cell injury, resveratrol prevents internucleosomal DNA fragmentation and apoptosis.[34] By monitoring the oxidation of dichlorofluorescein, resveratrol quenches the oxidative stress in Swiss 3T3 cells induced by the aldehydic product of the peroxidation of membrane w-6 polyunsaturated fatty acids, 4-HNE. In a follow-up study, we have shown that resveratrol prevents oxidave stress, apoptosis, and AP-1 activation induced by 4-HNE (Table 5.1).

The initial stages of arteriosclerosis involve changes in intracellular signaling cascades leading to endothelial dysfunction, secretion of vascular cell adhesion molecules, and smooth muscle cell proliferation contributing to the atherosclerotic plaque formation.[81–84] As mentioned before, the oxidation of extracellular LDL in the arterial wall is associated with endothelial dysfunction. The adhesion of circulating monocytes to dysfunctional endothelium is mediated by several adhesive molecules expressed on both monocyte and endothelial cell surface membranes. The adhesion molecules expressed on dysfunctional endothelial cells includes E-selectin, intracellular adhesion molecule-1 (ICAM-1), and vascular cell adhesion molecule-1 (VCAM-1).[71] Early observations by Ferrero et.al. suggest that resveratrol modifies ICAM-1 and VCAM-1 expression and induces a significant inhibition in the adhesion of monocytes on lipopolysaccharide-stimulated endothelial cells.[95] The genes of these endothelial cell adhesion factors contain NF-κβ transcription factor binding sites at their promoter regions.[71] Several studies within the last few years have shown that resveratrol inhibits NF-κβ-mediated gene expression.[96–100] Resveratrol mediates the suppression monocyte adhesion to endothelium by downregulating the E-selectin antigen expression on the endothelial surface, possibly through the NF-κβ pathway.[101] The increase of VCAM-1 mRNA and protein levels in bacterial lipopolysaccharide- or cytokine-treated human umbilical vein endothelial cells is prevented when cells are pretreated with resveratrol. This treatment remarkably reduces monocytoid cell adhesion to stimulated endothelium. Reporter gene assays reveal reduced activation of both NF-κβ and AP-1 after resveratrol pretreatment.[102] In neuronal cells, resveratrol protected cells from both the activation the NF-κβ/DNA binding activity and apoptotic cell death induced by oxidized lipoproteins.[103] The TNF- and LPS-induced activation of NF-κβ binding and subsequent apoptosis in human monocyte and myeloid cells is also diminished by resveratrol pretreatment.[96,97] Resveratrol also blocks the NF-κβ-dependent gene activation measured by a reduction in expression of monocyte chemoattractant protein-1, an NF-κβ-regulated gene product. The molecular mechanism of NF-κβ inactivation by resveratrol is not clear, but emerging evidence indicates the prevention of Iκβ phosphorylation and degradation is involved.[96,104,105]

Table 5.1 Effect of Resveratrol on 4-HNE Induced Oxidative Stress and Apoptosis

4-HNE
- Induces oxidative stress
- Induces apoptosis, activates caspase 3,9 and releases cyt c from mitochondria
- Upregulates AP-1; upregulates c-jun mRNA and protein levels
- Activates MAPK; upregulates p-JNK and p-p38 levels; downregulaates ERK 1/2

Resveratrol
- Acts as an antioxidant
- Prevents apoptosis; prevents caspase activation and cyt c-release
- Downregulates AP-1 activation; downregulates c-jun mRNA and protein levels
- Prevents MAPK activation; downregulates JNK and p-38; upregulates ERK 1/2

Proliferation and migration of vascular smooth muscle cells (VSMC) is one of the hallmarks of atherosclerosis. Red wine polyphenols hinder the proliferation and migration of cultured SMCs stimulated by growth factors through inhibition of PI3K activity and p38 (MAPK) phosphorylation.[106,107] In support, resveratrol reversibly arrests VSMC proliferation in early S phase of the cycle.[108] Recent studies on SMCs demonstrate the anti-atherogenic actions of resveratrol via its ability to inhibit the autocrine secretion and the mitogenic effects of endothelin-1.[109,110] Endothelin-1 (ET-1) is a small polypeptide with vasoconstrictive and mitogenic properties. ET-1 originates from endothelial cells and VSMCs, and induces the replacement of endothelial cells by neointimal SMC during arthrosclerotic plaque formation.[110,111] Induction of ET-1 expression in SMCs by H_2O_2-evoked oxidative stress is inhibited by resveratrol.[110] Moreover, resveratrol pretreatment suppresses ET-1-mediated protein tyrosine phosphorylation and MAPK activation, leading to SMCs proliferation and contraction.[109]

Another way ROS lead to impaired vasoregulation is the depletion of bioavailable NO and suppression of NO synthase (NOS) with subsequent ROS generation. NO is a potent vasodilator that also inhibits platelet and leukocyte adherence to vascular endothelium. Evidence suggests that resveratrol induces NOS in pulmonary endothelial cells, enhancing NO release and suppressing the proliferation of endothelial cells by arresting cell cycles.[112] In contrast, resveratrol inhibits NO generation and expression of cytosolic inducible NOS (iNOS) protein in LPS-stimulated macrophages by blocking NF-κβ activation.[100,113]

The cardiovascular protective effects rendered by resveratrol can also be ascribed to its ability to modulate arachidonic acid metabolism and to inhibit the platelet aggregation. Platelet aggregation is thought to be a precipitating event in cardiovascular disease, as platelet clumping may trigger heart attacks and strokes. Platelet aggregation is associated with the synthesis of eicosanoids from arachidonic acid. Initial observations indicate that resveratrol inhibits arachidonic acid-induced platelet aggregation and the production of proatherogenic eicosanoids by human platelets and neutrophils.[114,115] Resveratrol has been demonstrated to decrease platelet aggregation in human platelet-rich plasma by 50%.[116] More recently, it has been shown that LPS and phorbol ester treatment of peritoneal macrophages resulted in a COX-2 induction with subsequent increase in prostaglandin and arachidonic acid synthesis. Pretreatment of the cells with resveratrol prior to LPS or phorbol ester treatment exerted a strong inhibitory effect on COX-2 expression with a marked reduction of arachidonic acid and prostaglandin synthesis.[75] Activation of platelets by endotoxin or thrombin induces the adhesion of platelets to collagen and fibrinogen, the fibrillar constituent of atherosclerotic plaques. Preincubation of platelets with resveratrol inhibits this adhesion.[117]

In support of the antioxidant activity of resveratrol, a number of studies have demonstrated protective effects on oxidative cardiovascular injury.[118–122] In animal models, preinfusion of resveratrol prevents reperfusion-induced arrhythmias and mortality due to its antioxidant, free radical–scavenging activity as well as its ability to increase NO release.[123,124] These *in vivo* findings are supported by *in vitro* observations indicating that resveratrol significantly inhibits the ischemia/reperfusion-induced leukocyte recruitment and superoxide-related microvascular barrier dysfunction.[73] Furthermore, resveratrol induces the endogenous antioxidants and phase 2 enzymes in rat cardiomyocytes as part of a protective cellular defense against ROS and electrophilic species.[125]

CONCLUSION

The modulation of cellular signal transduction pathways by naturally occurring substances has been extended to understanding the molecular basis of atherosclerosis with dietary phytochemicals. Resveratrol has been shown to modulate diverse biochemical processes involved in atherosclerosis. Cellular signaling cascades mediated by oxidized lipids and their end products converge on AP-1 or NF-κβ, modulating the transcription of a number of genes. Results from our group as well as other laboratories suggest that resveratrol can act as an antioxidant by intercepting ROS-induced by-products of lipid peroxidation and prevent apoptosis. Furthermore, acting through MAPK-signaling pathways, resveratrol prevents AP-1 upregulation induced by 4-HNE and prevents activation of caspases and subsequent apoptosis. The overall picture suggests that the inhibitory effect of resveratrol on NF-κβ may turn off the genes induced by 4-HNE and turn on the genes initiating cell survival.

Although AP-1 or NF-κβ can be suggested as prime targets for any chemical or phytochemical, the stimuli and cell-specific cross talk between the transcription factors complicates the processes. Therefore, additional molecular studies will be needed to uncover the events taking place in the theme of "French paradox." Resveratrol undoubtedly warrants further study to more clearly elucidate its potential role in this diet-low disease risk relationship.

REFERENCES

1. Davis, R.J. 2000. Signal transduction by the JNK group of MAP kinases. *Cell* **103**:239–252.
2. Downward, J. 2001. The ins and outs of signalling. *Nature* **411**:759–762.
3. Hunter, T. 2000. Signaling — 2000 and beyond. *Cell* **100**:113–127.
4. Kossiakoff, A.A. 2004. The structural basis for biological signaling, regulation, and specificity in the growth hormone-prolactin system of hormones and receptors. *Adv. Protein Chem.* **68**:147–169.
5. Rollerova, E. and M. Urbancikova. 2000. Intracellular estrogen receptors, their characterization and function (Review). *Endocr. Regul.* **34**:203–218.
6. Aranda, A. and A. Pascual. 2001. Nuclear hormone receptors and gene expression. *Physiol. Rev.* **81**:1269–1304.
7. Jordan, J.D., E.M. Landau, and R. Iyengar. 2000. Signaling networks: the origins of cellular multitasking. *Cell* **103**:193–200.
8. Meeusen, T., I. Mertens, A. De Loof, and L. Schoofs. 2003. G protein-coupled receptors in invertebrates: a state of the art. *International Review of Cytology — A Survey of Cell Biology,* Vol 230 **230**:189–261.

9. Hulthe, J. and B. Fagerberg. 2002. Circulating oxidized LDL is associated with subclinical atherosclerosis development and inflammatory cytokines (AIR Study). *Arterioscler. Thromb. Vasc. Biol.* **22**:1162–1167.
10. O'Brien, K.D., C.E. Alpers, J.E. Hokanson, S. Wang, and A. Chait. 1996. Oxidation-specific epitopes in human coronary atherosclerosis are not limited to oxidized low-density lipoprotein. *Circulation* **94**:1216–1225.
11. Podrez, E.A., H.M. bu-Soud, and S.L. Hazen. 2000. Myeloperoxidase-generated oxidants and atherosclerosis. *Free Radic. Biol. Med.* **28**:1717–1725.
12. Negre-Salvayre, A., G. Fitoussi, V. Reaud, M.T. Pieraggi, J.C. Thiers, and R. Salvayre. 1992. A delayed and sustained rise of cytosolic calcium is elicited by oxidized LDL in cultured bovine aortic endothelial cells. *FEBS Lett.* **299**:60–65.
13. Weisser, B., R. Locher, T. Mengden, and W. Vetter. 1992. Oxidation of low density lipoprotein enhances its potential to increase intracellular free calcium concentration in vascular smooth muscle cells. *Arterioscler. Thromb.* **12**:231–236.
14. Fyrnys, B., R. Claus, G. Wolf, and H.P. Deigner. 1997. Oxidized low density lipoprotein stimulates protein kinase C (PKC) activity and expression of PKC-isotypes via prostaglandin-H-synthase in P388D1 cells. *Adv. Exp. Med. Biol.* **407**:93–98.
15. Suc, I., O. Meilhac, I. Lajoie-Mazenc, J. Vandaele, G. Jurgens, R. Salvayre, and A. Negre-Salvayre. 1998. Activation of EGF receptor by oxidized LDL. *FASEB J.* **12**:665–671.
16. Khan, B.V., S.S. Parthasarathy, R.W. Alexander, and R.M. Medford. 1995. Modified low density lipoprotein and its constituents augment cytokine-activated vascular cell-adhesion molecule-1 gene-expression in human vascular endothelial-cells. *J. of Clin. Invest.* **95**:1262–1270.
17. Kamanna, V.S., R. Pai, H. Ha, M.A. Kirschenbaum, and D.D. Roh. 1999. Oxidized low-density lipoprotein stimulates monocyte adhesion to glomerular endothelial cells. *Kidney Int.* **55**:2192–2202.
18. Weber, C., W. Erl, K.S. Weber, and P.C. Weber. 1999. Effects of oxidized low density lipoprotein, lipid mediators and statins on vascular cell interactions. *Clin. Chem. Lab Med.* **37**:243–251.
19. Maziere, C., M. Djavaheri–Mergny, V. FreyFressart, J. Delattre, and J.C. Maziere. 1997. Copper and cell-oxidized low-density lipoprotein induces activator protein 1 in fibroblasts, endothelial and smooth muscle cells. *FEBS Lett.* **409**:351–356.
20. Maziere, C., M. Auclair, M. Djavaheri-Mergny, L. Packer, and J.C. Maziere. 1996. Oxidized low density lipoprotein induces activation of the transcription factor NF kappa B in fibroblasts, endothelial and smooth muscle cells. *Biochem. Mol. Biol. Int.* **39**:1201–1207.
21. Nagy, L., P. Tontonoz, J.G. Alvarez, H. Chen, and R.M. Evans. 1998. Oxidized LDL regulates macrophage gene expression through ligand activation of PPARgamma. *Cell* **93**:229–240.
22. Tontonoz, P., L. Nagy, J.G. Alvarez, V.A. Thomazy, and R.M. Evans. 1998. PPAR-gamma promotes monocyte/macrophage differentiation and uptake of oxidized LDL. *Cell* **93**:241–252.
23. Ross, R. 1979. The pathogenesis of atherosclerosis. *Mech. Ageing Dev.* **9**:435–440.
24. Ross, R. 1995. Cell biology of atherosclerosis. *Annu. Rev. Physiol.* **57**:791–804.
25. Brown, A.J., S.L. Leong, R.T. Dean, and W. Jessup. 1997. 7-Hydroperoxycholesterol and its products in oxidized low density lipoprotein and human atherosclerotic plaque. *J. Lipid Res.* **38**:1730–1745.

26. Hodis, H.N., D.W. Crawford, and A. Sevanian. 1991. Cholesterol feeding increases plasma and aortic tissue cholesterol oxide levels in parallel: further evidence for the role of cholesterol oxidation in atherosclerosis. *Atherosclerosis* **89**:117–126.

27. Kraemer, R., K.B. Pomerantz, J. Joseph-Silverstein, and D.P. Hajjar. 1993. Induction of basic fibroblast growth factor mRNA and protein synthesis in smooth muscle cells by cholesteryl ester enrichment and 25-hydroxycholesterol. *J. Biol. Chem.* **268**:8040–8045.

28. Esterbauer, H., R.J. Schaur, and H. Zollner. 1991. Chemistry and biochemistry of 4-hydroxynonenal, malonaldehyde and related aldehydes. *Free Radic. Biol. Med.* **11**:81–128.

29. Dianzani, M.U. 1998. 4-Hydroxynonenal and cell signalling. *Free Radic. Res.* **28**:553–560.

30. Sayre, L.M., D.A. Zelasko, P.L. Harris, G. Perry, R.G. Salomon, and M.A. Smith. 1997. 4-Hydroxynonenal-derived advanced lipid peroxidation end products are increased in Alzheimer's disease. *J. Neurochem.* **68**:2092–2097.

31. Paradis, V., M. Kollinger, M. Fabre, A. Holstege, T. Poynard, and P. Bedossa. 1997. *In situ* detection of lipid peroxidation by-products in chronic liver diseases. *Hepatology* **26**:135–142.

32. Smith, R.G., Y.K. Henry, M.P. Mattson, and S.H. Appel. 1998. Presence of 4-hydroxynonenal in cerebrospinal fluid of patients with sporadic amyotrophic lateral sclerosis. *Ann. Neurol.* **44**:696–699.

33. Napoli, C., F.P. D'Armiento, F.P. Mancini, A. Postiglione, J.L. Witztum, G. Palumbo, and W. Palinski. 1997. Fatty streak formation occurs in human fetal aortas and is greatly enhanced by maternal hypercholesterolemia. Intimal accumulation of low density lipoprotein and its oxidation precede monocyte recruitment into early atherosclerotic lesions. *J. Clin. Invest.* **100**:2680–2690.

34. Kutuk, O., M. Adli, G. Poli, and H. Basaga. 2004. Resveratrol protects against 4-HNE induced oxidative stress and apoptosis in Swiss 3T3 fibroblasts. *Biofactors* **20**:1–10.

35. Parola, M., R. Muraca, I. Dianzani, G. Barrera, G. Leonarduzzi, P. Bendinelli, R. Piccoletti, and G. Poli. 1992. Vitamin E dietary supplementation inhibits transforming growth factor beta 1 gene expression in the rat liver. *FEBS Lett.* **308**:267–270.

36. Leonarduzzi, G., A. Scavazza, F. Biasi, E. Chiarpotto, S. Camandola, S. Vogel, R. Dargel, and G. Poli. 1997. The lipid peroxidation end product 4-hydroxy-2,3-nonenal up-regulates transforming growth factor beta1 expression in the macrophage lineage: a link between oxidative injury and fibrosclerosis. *FASEB J.* **11**:851–857.

37. Parola, M., M. Pinzani, A. Casini, E. Albano, G. Poli, A. Gentilini, P. Gentilini, and M.U. Dianzani. 1993. Stimulation of lipid peroxidation or 4-hydroxynonenal treatment increases procollagen alpha 1 (I) gene expression in human liver fat-storing cells. *Biochem. Biophys. Res. Commun.* **194**:1044–1050.

38. Kim, S.J., F. Denhez, K.Y. Kim, J.T. Holt, M.B. Sporn, and A.B. Roberts. 1989. Activation of the second promoter of the transforming growth factor-beta 1 gene by transforming growth factor-beta 1 and phorbol ester occurs through the same target sequences. *J. Biol. Chem.* **264**:19373–19378.

39. Armendariz-Borunda, J., C.P. Simkevich, N. Roy, R. Raghow, A.H. Kang, and J.M. Seyer. 1994. Activation of Ito cells involves regulation of AP-1 binding proteins and induction of type I collagen gene expression. *Biochem. J.* **304** (**Pt 3**):817–824.

40. Camandola, S., A. Scavazza, G. Leonarduzzi, F. Biasi, E. Chiarpotto, A. Azzi, and G. Poli. 1997. Biogenic 4-hydroxy-2-nonenal activates transcription factor AP-1 but not NF-kappa B in cells of the macrophage lineage. *Biofactors* **6**:173–179.

41. Parola, M., G. Robino, F. Marra, M. Pinzani, G. Bellomo, G. Leonarduzzi, P. Chiarugi, S. Camandola, G. Poli, G. Waeg, P. Gentilini, and M.U. Dianzani. 1998. HNE interacts directly with JNK isoforms in human hepatic stellate cells. *J. Clin. Invest.* **102**:1942–1950.

42. Camandola, S., G. Poli, and M.P. Mattson. 2000. The lipid peroxidation product 4-hydroxy-2,3-nonenal increases AP-1-binding activity through caspase activation in neurons. *J. Neurochem.* **74**:159–168.

43. Page, S., C. Fischer, B. Baumgartner, M. Haas, U. Kreusel, G. Loidl, M. Hayn, H.W. Ziegler-Heitbrock, D. Neumeier, and K. Brand. 1999. 4-Hydroxynonenal prevents NF-kappaB activation and tumor necrosis factor expression by inhibiting IkappaB phosphorylation and subsequent proteolysis. *J. Biol. Chem.* **274**:11611–11618.

44. Uchida, K., M. Shiraishi, Y. Naito, Y. Torii, Y. Nakamura, and T. Osawa. 1999. Activation of stress signaling pathways by the end product of lipid peroxidation. 4-hydroxy-2-nonenal is a potential inducer of intracellular peroxide production. *J. Biol. Chem.* **274**:2234–2242.

45. Ruef, J., G.N. Rao, F. Li, C. Bode, C. Patterson, A. Bhatnagar, and M.S. Runge. 1998. Induction of rat aortic smooth muscle cell growth by the lipid peroxidation product 4-hydroxy-2-nonenal. *Circulation* **97**:1071–1078.

46. Chiarpotto, E., C. Domenicotti, D. Paola, A. Vitali, M. Nitti, M.A. Pronzato, F. Biasi, D. Cottalasso, U.M. Marinari, A. Dragonetti, P. Cesaro, C. Isidoro, and G. Poli. 1999. Regulation of rat hepatocyte protein kinase C beta isoenzymes by the lipid peroxidation product 4-hydroxy-2,3-nonenal: A signaling pathway to modulate vesicular transport of glycoproteins. *Hepatology* **29**:1565–1572.

47. Cornwell, T., W. Cohick, and I. Raskin. 2004. Dietary phytoestrogens and health. *Phytochemistry* **65**:995–1016.

48. Oak, M.H., B.J. El, and V.B. Schini-Kerth. 2005. Antiangiogenic properties of natural polyphenols from red wine and green tea. *J. Nutr. Biochem.* **16**:1–8.

49. Park, E.J. and J.M. Pezzuto. 2002. Botanicals in cancer chemoprevention. *Cancer Metastasis Rev.* **21**:231–255.

50. Lee, H.J., J.W. Seo, B.H. Lee, K.H. Chung, and D.Y. Chi. 2004. Syntheses and radical scavenging activities of resveratrol derivatives. *Bioorg. Med. Chem. Lett.* **14**:463–466.

51. Jia, W., W.Y. Gao, Y. Q. Yan, J. Wang, Z.H. Xu, W.J. Zheng, and P.G. Xiao. 2004. The rediscovery of ancient Chinese herbal formulas. *Phytother. Res.* **18**:681–686.

52. Sumner, L.W., P. Mendes, and R.A. Dixon. 2003. Plant metabolomics: large-scale phytochemistry in the functional genomics era. *Phytochemistry* **62**:817–836.

53. Dixon, R.A. and D. Ferreira. 2002. Genistein. *Phytochemistry* **60**:205–211.

54. Mazur, W.M., K. Wahala, S. Rasku, A. Salakka, T. Hase, and H. Adlercreutz. 1998. Lignan and isoflavonoid concentrations in tea and coffee. *Br. J. Nutr.* **79**:37–45.

55. Glitso, L.V., W.M. Mazur, H. Adlercreutz, K. Wahala, T. Makela, B. Sandstrom, and K.E. Bach Knudsen. 2000. Intestinal metabolism of rye lignans in pigs. *Br. J. Nutr.* **84**:429–437.

56. Setchell, K.D., A.M. Lawson, S.P. Borriello, R. Harkness, H. Gordon, D.M. Morgan, D.N. Kirk, H. Adlercreatz, L.C. Anderson, and M. Axelson. 1981. Lignan formation in man — microbial involvement and possible roles in relation to cancer. *Lancet* **2**:4–7.

57. Nesbitt, P.D. and L.U. Thompson. 1997. Lignans in homemade and commercial products containing flaxseed. *Nutr. Cancer* **29**:222–227.

58. Franke, A.A., L.J. Custer, C.M. Cerna, and K. Narala. 1995. Rapid HPLC analysis of dietary phytoestrogens from legumes and from human urine. *Proc. Soc. Exp. Biol. Med.* **208**:18–26.

59. Knuckles, B.E., D. deFremery, and G.O. Kohler. 1976. Coumestrol content of fractions obtained during wet processing of alfalfa. *J. Agric. Food Chem.* **24**:1177–1180.
60. Gehm, B.D., J.M. McAndrews, P.Y. Chien, and J.L. Jameson. 1997. Resveratrol, a polyphenolic compound found in grapes and wine, is an agonist for the estrogen receptor. *Proc. Natl. Acad. Sci. USA* **94**:14138–14143.
61. Burr, M.L. 1995. Explaining the French paradox. *J.R. Soc. Health* **115**:217–219.
62. Cleophas, T.J., E. Tuinenberg, M.J. van der, and K.H. Zwinderman. 1996. Wine consumption and other dietary variables in males under 60 before and after acute myocardial infarction. *Angiology* **47**:789–796.
63. Drewnowski, A., S.A. Henderson, A.B. Shore, C. Fischler, P. Preziosi, and S. Hercberg. 1996. Diet quality and dietary diversity in France: implications for the French paradox. *J. Am. Diet. Assoc.* **96**:663–669.
64. Law, M. and N. Wald. 1999. Why heart disease mortality is low in France: the time lag explanation. *BMJ* **318**:1471–1476.
65. Renaud, S. and J.C. Ruf. 1994. The French paradox: vegetables or wine. *Circulation* **90**:3118–3119.
66. Fremont, L. 2000. Biological effects of resveratrol. *Life Sci.* **66**:663–673.
67. Burns, J., T. Yokota, H. Ashihara, M.E. Lean, and A. Crozier. 2002. Plant foods and herbal sources of resveratrol. *J. Agric. Food Chem.* **50**:3337–3340.
68. Chen, R.S., P.L. Wu, and R.Y. Chiou. 2002. Peanut roots as a source of resveratrol. *J. Agric. Food Chem.* **50**:1665–1667.
69. Bertelli, A.A., L. Giovannini, R. Stradi, S. Urien, J.P. Tillement, and A. Bertelli. 1996. Kinetics of trans- and cis-resveratrol (3,4',5-trihydroxystilbene) after red wine oral administration in rats. *Int. J. Clin. Pharmacol. Res.* **16**:77–81.
70. Jannin, B., M. Menzel, J.P. Berlot, D. Delmas, A. Lancon, and N. Latruffe. 2004. Transport of resveratrol, a cancer chemopreventive agent, to cellular targets: plasmatic protein binding and cell uptake. *Biochem. Pharmacol.* **68**:1113–1118.
71. Kutuk, O. and H. Basaga. 2003. Inflammation meets oxidation: NF-kappaB as a mediator of initial lesion development in atherosclerosis. *Trends Mol. Med.* **9**:549–557.
72. Bonnefoy, M., J. Drai, and T. Kostka. 2002. Antioxidants to slow aging, facts and perspectives. *Presse Med.* **31**:1174–1184.
73. Shigematsu, S., S. Ishida, M. Hara, N. Takahashi, H. Yoshimatsu, T. Sakata, and R.J. Korthuis. 2003. Resveratrol, a red wine constituent polyphenol, prevents superoxide-dependent inflammatory responses induced by ischemia/reperfusion, platelet-activating factor, or oxidants. *Free Radic. Biol. Med.* **34**:810–817.
74. Karlsson, J., M. Emgard, P. Brundin, and M.J. Burkitt. 2000. trans-resveratrol protects embryonic mesencephalic cells from tert-butyl hydroperoxide: electron paramagnetic resonance spin trapping evidence for a radical scavenging mechanism. *J. Neurochem.* **75**:141–150.
75. Martinez, J. and J.J. Moreno. 2000. Effect of resveratrol, a natural polyphenolic compound, on reactive oxygen species and prostaglandin production. *Biochem. Pharmacol.* **59**:865–870.
76. Leonard, S.S., C. Xia, B.H. Jiang, B. Stinefelt, H. Klandorf, G.K. Harris, and X. Shi. 2003. Resveratrol scavenges reactive oxygen species and effects radical-induced cellular responses. Biochem. *Biophys. Res. Commun.* **309**:1017–1026.
77. Esterbauer, H., M. Dieber-Rotheneder, G. Waeg, G. Striegl, and G. Jurgens. 1990. Biochemical, structural, and functional properties of oxidized low-density lipoprotein. *Chem. Res. Toxicol.* **3**:77–92.

78. Kalant, N., S. McCormick, and M.A. Parniak. 1991. Effects of copper and histidine on oxidative modification of low density lipoprotein and its subsequent binding to collagen. *Arterioscler. Thromb.* **11**:1322–1329.

79. Morris, C.J., G.V. Bradby, and K.W. Walton. 1978. Fibrous long-spacing collagen in human atherosclerosis. *Atherosclerosis* **31**:345–354.

80. Simonini, G., A. Pignone, S. Generini, F. Falcini, and M.M. Cerinic. 2000. Emerging potentials for an antioxidant therapy as a new approach to the treatment of systemic sclerosis. *Toxicology* **155**:1–15.

81. Dong, Z.M. and D.D. Wagner. 1998. Leukocyte-endothelium adhesion molecules in atherosclerosis. *J. Lab Clin. Med.* **132**:369–375.

82. Kaplan, M. and M. Aviram. 2001. Retention of oxidized LDL by extracellular matrix proteoglycans leads to its uptake by macrophages: an alternative approach to study lipoproteins cellular uptake. *Arterioscler. Thromb. Vasc. Biol.* **21**:386–393.

83. Simionescu, M. and N. Simionescu. 1993. Proatherosclerotic events: pathobiochemical changes occurring in the arterial wall before monocyte migration. *FASEB J.* **7**:1359–1366.

84. Skalen, K., M. Gustafsson, E.K. Rydberg, L.M. Hulten, O. Wiklund, T.L. Innerarity, and J. Boren. 2002. Subendothelial retention of atherogenic lipoproteins in early atherosclerosis. *Nature* **417**:750–754.

85. Hertog, M.G., E.J. Feskens, and D. Kromhout. 1997. Antioxidant flavonols and coronary heart disease risk. *Lancet* **349**:699.

86. Vinson, J.A., K. Teufel, and N. Wu. 2001. Red wine, dealcoholized red wine, and especially grape juice, inhibit atherosclerosis in a hamster model. *Atherosclerosis* **156**:67–72.

87. Miura, D., Y. Miura, and K. Yagasaki. 2003. Hypolipidemic action of dietary resveratrol, a phytoalexin in grapes and red wine, in hepatoma-bearing rats. *Life Sci.* **73**:1393–1400.

88. Filip, V., M. Plockova, J. Smidrkal, Z. Spickova, K. Melzoch, and S. Schmidt. 2003. Resveratrol and its antioxidant and antimicrobial effectiveness. *Food Chemistry* **83**:585–593.

89. Belguendouz, L., L. Fremont, and A. Linard. 1997. Resveratrol inhibits metal ion-dependent and independent peroxidation of porcine low-density lipoproteins. *Biochem. Pharmacol.* **53**:1347–1355.

90. Fremont, L., L. Belguendouz, and S. Delpal. 1999. Antioxidant activity of resveratrol and alcohol-free wine polyphenols related to LDL oxidation and polyunsaturated fatty acids. *Life Sciences* **64**:2511–2521.

91. Sun, A.Y., Y.M. Chen, M. JamesKracke, P. Wixom, and Y. Cheng. 1997. Ethanol-induced cell death by lipid peroxidation in PC12 cells. *Neurochem. Res.* **22**:1187–1192.

92. Stivala, L.A., M. Savio, F. Carafoli, P. Perucca, L. Bianchi, G. Maga, L. Forti, U.M. Pagnoni, A. Albini, E. Prosperi, and V. Vannini. 2001. Specific structural determinants are responsible for the antioxidant activity and the cell cycle effects of resveratrol. *J. Biol. Chem.* **276**:22586–22594.

93. Jang, J.H. and Y.J. Surh. 2001. Protective effects of resveratrol on hydrogen peroxide-induced apoptosis in rat pheochromocytoma (PC12) cells. *Mutation Research-Genetic Toxicology and Environmental Mutagenesis* **496**:181–190.

94. Jang, J.H. and Y.J. Surh. 2003. Protective effect of resveratrol on beta-amyloid-induced oxidative PC12 cell death. *Free Radic. Biol. Med.* **34**:1100–1110.

95. Ferrero, M.E., A.A. E. Bertelli, A. Fulgenzi, F. Pellegatta, M.M. Corsi, M. Bonfrate, F. Ferrara, R. De Caterina, L. Giovannini, and A. Bertelli. 1998. Activity in vitro of resveratrol on granulocyte and monocyte adhesion to endothelium. *Am. J. of Clin. Nutr.* **68**:1208–1214.

96. Holmes-McNary, M. and A.S. Baldwin. 2000. Chemopreventive properties of trans-resveratrol are associated with inhibition of activation of the I kappa B kinase. *Cancer Res.* **60**:3477–3483.

97. Manna, S.K., A. Mukhopadhyay, and B.B. Aggarwal. 2000. Resveratrol suppresses TNF-induced activation of nuclear transcription factors NF-kappa B, activator protein-1, and apoptosis: Potential role of reactive oxygen intermediates and lipid peroxidation. *J. Immunol.* **164**:6509–6519.

98. Pendurthi, U.R., J.T. Williams, and L.V. M. Rao. 1999. Resveratrol, a polyphenolic compound found in wine, inhibits tissue factor expression in vascular cells — a possible mechanism for the cardiovascular benefits associated with moderate consumption of wine. *Arterioscler. Thromb. Vasc. Biol.* **19**:419–426.

99. Pendurthi, U.R. and L.V.M. Rao. 2002. Resveratrol suppresses agonist-induced monocyte adhesion to cultured human endothelial cells. *Thromb. Res.* **106**:243–248.

100. Tsai, S.H., S.Y. Lin-Shiau, and J.K. Lin. 1999. Suppression of nitric oxide synthase and the down-regulation of the activation of NF kappa B in macrophages by resveratrol. *Br. J. Pharmacol.* **126**:673–680.

101. Pendurthi, U.R., F. Meng, N. Mackman, and L.V.M. Rao. 2002. Mechanism of resveratrol-mediated suppression of tissue factor gene expression. *Thrombosis and Haemostasis* **87**:155–162.

102. Carluccio, M.A., L. Siculella, M.A. Ancora, M. Massaro, E. Scoditti, C. Storelli, F. Visioli, A. Distante, and R. De Caterina. 2003. Olive oil and red wine antioxidant polyphenols inhibit endothelial activation — antiatherogenic properties of Mediterranean diet phytochemicals. *Arterioscler. Thromb. Vasc. Biol.* **23**:622–629.

103. Draczynska-Lusiak, B., Y.M. Chen, and A.Y. Sun. 1998. Oxidized lipoproteins activate NF-kappa B binding activity and apoptosis in PC12 cells. *Neuroreport* **9**:527–532.

104. Murakami, A., K. Matsumoto, K. Koshimizu, and H. Ohigashi. 2003. Effects of selected food factors with chemopreventive properties on combined lipopolysaccharide — and interferon-gamma-induced I kappa B degradation in RAW264.7 macrophages. *Cancer Lett.* **195**:17–25.

105. Storz, P., H. Doppler, and A. Toker. 2004. Activation loop phosphorylation controls protein kinase D-dependent activation of nuclear factor kappa B. *Mol. Pharmacol.* **66**:870–879.

106. Araim, O., J. Ballantyne, A.L. Waterhouse, and B.E. Sumpio. 2002. Inhibition of vascular smooth muscle cell proliferation with red wine and red wine polyphenols. *J. Vasc. Surg.* **35**:1226–1232.

107. Iijima, K., M. Yoshizumi, M. Hashimoto, M. Akishita, K. Kozaki, J. Ako, T. Watanabe, Y. Ohike, B. Son, J. Yu, K. Nakahara, and Y. Ouchi. 2002. Red wine polyphenols inhibit vascular smooth muscle cell migration through two distinct signaling pathways. *Circulation* **105**:2404–2410.

108. Haider, U.G., D. Sorescu, K.K. Griendling, A.M. Vollmar, and V.M. Dirsch. 2003. Resveratrol increases serine15-phosphorylated but transcriptionally impaired p53 and induces a reversible DNA replication block in serum-activated vascular smooth muscle cells. *Mol. Pharmacol.* **63**:925–932.

109. El-Mowafy, A.M. and R.E. White. 1999. Resveratrol inhibits MAPK activity and nuclear translocation in coronary artery smooth muscle: reversal of endothelin-1 stimulatory effects. *FEBS Lett.* **451**:63–67.

110. Ruef, J., M. Moser, W. Kubler, and C. Bode. 2001. Induction of endothelin-1 expression by oxidative stress in vascular smooth muscle cells. *Cardiovasc. Pathol.* **10**:311–315.

111. Raines, E.W. and R. Ross. 1993. Smooth muscle cells and the pathogenesis of the lesions of atherosclerosis. *Br. Heart J.* **69**:S30–S37.

112. Hsieh, T.C., G. Juan, Z. Darzynkiewicz, and J.M. Wu. 1999. Resveratrol increases nitric oxide synthase, induces accumulation of p53 and p21(WAF1/CIP1), and suppresses cultured bovine pulmonary artery endothelial cell proliferation by perturbing progression through S and G2. *Cancer Res.* **59**:2596–2601.

113. Cho, D.I., N.Y. Koo, W.J. Chung, T.S. Kim, S.Y. Ryu, S.Y. Im, and K.M. Kim. 2002. Effects of resveratrol-related hydroxystilbenes on the nitric oxide production in macrophage cells: structural requirements and mechanism of action. *Life Sciences* **71**:2071–2082.

114. PaceAsciak, C.R., O. Rounova, S.E. Hahn, E.P. Diamandis, and D.M. Goldberg. 1996. Wines and grape juices as modulators of platelet aggregation in healthy human subjects. *Clinica Chimica Acta* **246**:163–182.

115. Subbaramaiah, K., W.J. Chung, P. Michaluart, N. Telang, T. Tanabe, H. Inoue, M.S. Jang, J.M. Pezzuto, and A.J. Dannenberg. 1998. Resveratrol inhibits cyclooxygenase-2 transcription and activity in phorbol ester-treated human mammary epithelial cells. *J. Biol. Chem.* **273**:21875–21882.

116. Bertelli, A.A., L. Giovannini, D. Giannessi, M. Migliori, W. Bernini, M. Fregoni, and A. Bertelli. 1995. Antiplatelet activity of synthetic and natural resveratrol in red wine. *Int. J. Tissue React.* **17**:1–3.

117. Olas, B., B. Wachowicz, J. Saluk-Juszczak, and T. Zielinski. 2002. Effect of resveratrol, a natural polyphenolic compound, on platelet activation induced by endotoxin or thrombin. *Thromb. Res.* **107**:141–145.

118. Cuzzocrea, S., D.P. Riley, A.P. Caputi, and D. Salvemini. 2001. Antioxidant therapy: A new pharmacological approach in shock, inflammation, and ischemia/reperfusion injury. *Pharmacol. Rev.* **53**:135–159.

119. Hattori, R., H. Otani, N. Maulik, and D.K. Das. 2002. Pharmacological preconditioning with resveratrol: role of nitric oxide. *Am. J. Physiol. Heart Circ. Physiol.* **282**:H1988–H1995.

120. Lefer, D.J. and D.N. Granger. 2000. Oxidative stress and cardiac disease. *Am. J. Med.* **109**:315–323.

121. Marchioli, R. 1999. Antioxidant vitamins and prevention of cardiovascular disease: Laboratory, epidemiological and clinical trial data. *Pharmacol. Res.* **40**:227–238.

122. Wu, J.M., Z.R. Wang, T.C. Hsieh, J.L. Bruder, J.G. Zou, and Y.Z. Huang. 2001. Mechanism of cardioprotection by resveratrol, a phenolic antioxidant present in red wine (Review). *Int. J. Mol. Med.* **8**:3–17.

123. Hung, L.M., J.K. Chen, S.S. Huang, R.S. Lee, and M.J. Su. 2000. Cardioprotective effect of resveratrol, a natural antioxidant derived from grapes. *Cardiovasc. Res.* **47**:549–555.

124. Hung, L.M., M.J. Su, and J.K. Chen. 2004. Resveratrol protects myocardial ischemia-reperfusion injury through both NO-dependent and NO-independent mechanisms. *Free Radic. Biol. Med.* **36**:774–781.

125. Cao, Z. and Y. Li. 2004. Potent induction of cellular antioxidants and phase 2 enzymes by resveratrol in cardiomyocytes: protection against oxidative and electrophilic injury. *Eur. J. Pharmacol.* **489**:39–48.

CHAPTER **6**

Adipose Tissue Gene Expression in the Context of Inflammation and Obesity

Philip A. Kern

CONTENTS

INTRODUCTION

Obesity is among the most common diseases facing physicians. It is a major cause of premature mortality, with enormous economic impact on our society. A recent study estimated the annual cost of obesity in the U.S. as $93 billion per year, or 9.1% of all healthcare dollars, of which about half is paid by federal dollars.[1] As frightening as these numbers are, they are based on 1998 data on the prevalence of obesity, and recent studies have indicated that the obesity epidemic has continued unabated,[2] and the percent of adults with BMI >30 kg/m^2 has increased from 18.3% to 22.5% in the 4 years from 1998 to 2002 (CDC, Behavioral Risk Factor Surveillance System).

This paper will provide an overview of the pathogenesis of obesity, with particular emphasis on metabolic syndrome and high risk obesity. In this review, we present evidence to suggest that the accumulation of adipose tissue, per se, is not necessarily a precursor to the metabolic syndrome that accompanies obesity, but the accumulation of lipid "in all the wrong places" leads to many of the adverse outcomes.[3]

CLINICAL FEATURES OF OBESITY, METABOLIC SYNDROME, AND TYPE 2 DIABETES

Although obesity is defined as an excess of adipose tissue, a useful classification of obesity is based not on adiposity alone but as excess adipose tissue leading to a spectrum of health consequences. These adverse effects on health usually occur when a patient exceeds "normal" body weight by 20% or more. However, there is considerable variation among individuals, and some subjects can become remarkably obese with few, if any, metabolic consequences, whereas others develop metabolic syndrome with minimal weight gain. There are numerous approaches to measuring degrees of obesity, and the body mass index (BMI) has gained widespread acceptance in a clinical setting. The BMI represents weight (kg) divided by height (m)^2. Although the terminology is imprecise, the normal weight range usually falls between a BMI of 20 and 25 kg/m^2. A BMI range of 25 to 30 kg/m^2 is often referred to as "overweight, and a BMI >30 kg/m^2 is obese. Severe obesity is characterized by a BMI in excess of 35.

In population studies, most of the complications of obesity increase with progressive increases in BMI, beginning when the BMI exceeds 25 kg/m^2.[4] For example, the relative risk of type 2 diabetes in relation to BMI is shown in Figure 6.1 using data from the Nurses Health Study.[5] The risk of developing diabetes begins to increase at a BMI of 25, however this risk is three-fold higher when the BMI reaches

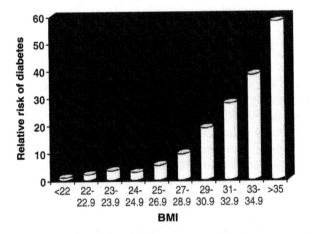

Figure 6.1   BMI and the relative risk of type 2 diabetes from the Women's Health Study. The relative risk for the development of type 2 diabetes over 8 years for women 30–55 years of age in 1976 is shown. The data are adjusted for age, and the risk of diabetes at a BMI <22 kg/m^2 is adjusted to 1.0. (Data from Colditz et al., *Am. J. Epidemiol.*, 132:501–513. Redrawn with permission, Oxford University Press.)

30. This increasing risk of diabetes with obesity is particularly disturbing, since the prevalence of obesity has increased in an epidemic fashion over the last 20 years. It is therefore not surprising that the prevalence of type 2 diabetes and impaired glucose tolerance has risen during this same time period.[6,7]

In addition to diabetes, there are many health risks associated with obesity, ranging from metabolic disturbances such as hyperlipidemia, insulin resistance, and hypertension, to sleep apnea, gallstones, arthritis, and an increased risk for several malignancies.[8,9] These obesity-related medical problems are compounded by the social stigmata of obesity, depression, and the difficulties that obesity poses on the physical examination, diagnostic medical tests, and otherwise routine surgery.

The constellation of insulin resistance, abnormal lipid metabolism, hypertension, and abdominal fat distribution are genetically linked and constitute the metabolic syndrome (Table 6.1). Body fat distribution is a particularly important factor in determining high risk obesity. Patients with abdominal obesity are at greater risk for heart disease, diabetes, hypertension, and hyperlipidemia compared to patients with more gluteal fat distribution.[10] In a recent study, the removal of subcutaneous fat by liposuction, without any removal of visceral fat, resulted in no improvement in insulin resistance or inflammation, further highlighting the importance of the visceral adipose depot.[11] This difference in risk level with adipose tissue distribution is especially important for mildly obese persons. The reason for the abdominal obesity-metabolic syndrome association is not clear, but one leading concept maintains that visceral adipose tissue has a higher rate of lipolysis, resulting in elevated portal non-esterified fatty acids, which increase hepatic VLDL production, increase hepatic glucose production, and impair peripheral insulin sensitivity.[12]

The development of type 2 diabetes coincides with progressive obesity but occurs in individuals who are genetically susceptible. The risks of diabetes include microvascular disease, resulting in diabetic retinopathy and nephropathy, along with a number of different forms of neuropathy, and accelerated systemic atherosclerosis, resulting in coronary artery, cerebrovascular, and peripheral vascular disease. The atherosclerosis of diabetes is complex, but is not entirely dependent on hyperglycemia since patients with insulin resistance, hyperinsulinemia, and the metabolic syndrome are also at increased risk for atherosclerosis. There is a strong statistical association between obesity and metabolic syndrome, and most patients will demonstrate some features of metabolic syndrome with more severe obesity. A minority

Table 6.1 Features of the Metabolic Syndrome

Abnormal carbohydrate metabolism: type 2 diabetes, impaired glucose tolerance/impaired fasting glucose, insulin resistance, hyperinsulinemia
Hypertriglyceridemia
Low HDL
Hypertension
Increased visceral adipose tissue

of patients, however, are remarkably immune to the metabolic consequences of obesity, in spite of extreme obesity, whereas other patients develop the full spectrum of metabolic syndrome with only a small incremental increase in BMI.

ECTOPIC LIPID ACCUMULATION AND METABOLIC SYNDROME

Adipose tissue is the primary organ of lipid storage. However, with progressive obesity, lipid is deposited into other non-adipose organs, provoking the term "ectopic lipid" (Figure 6.2). Some of this ectopic lipid is deposited in the liver, leading to the commonly observed fatty liver, also known as non-alcoholic steatohepatitis (NASH), or hepatic steatosis. Fatty liver is very common in patients with metabolic syndrome, and is associated with elevated glucose production by the liver, which is one of the hallmarks of type 2 diabetes. NASH may also lead to the development of hepatic cirrhosis, although the precise pathogenesis of NASH-mediated cirrhosis is not known.

Subjects who are obese or who have metabolic syndrome usually demonstrate hyperinsulinemia.[13] Although this hyperinsulinemia might be interpreted initially as robust β-cell function, when one considers the degree of insulin resistance, β-cell compensation shows signs of deterioration long before the development of type 2 diabetes.[14–16] Much of this deterioration in β-cell function is likely due to lipid accumulation in islets. Although this has not been demonstrated in humans, Zucker diabetic fatty rats demonstrated islet lipid accumulation, apoptosis, and decreased cell mass and function.[17]

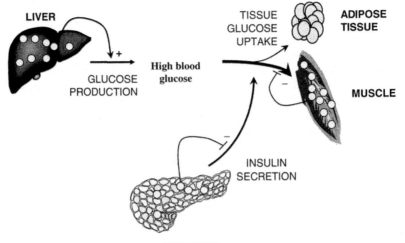

Figure 6.2 Ectopic lipid accumulation and insulin resistance. As obesity progresses, lipid accumulation occurs in other organs ("ectopic" sites). The lipid accumulation in these ectopic sites leads to insulin resistance (impaired glucose uptake in muscle, and increased glucose output in liver), and to impaired insulin secretion from the β-cells of the pancreas.

Most of peripheral glucose disposal occurs in skeletal muscle,[18] making this tissue extremely important in understanding the mechanisms underlying peripheral insulin resistance. Insulin resistant subjects demonstrate increased intramyocellular lipid (IMCL), decreased proportion of the oxidative type I and type IIA muscle fibers, decreased oxidative capacity of each class of muscle fiber, and decreased capillary density.[19,20] The cause-effect relationship between muscle glucose uptake and muscle lipid metabolism is not well understood. However, there is accumulating evidence that IMCL induces lipotoxicity, which affects peripheral insulin sensitivity and muscle glucose transport.[21-24] In lean subjects who are at risk for the development of type 2 diabetes, there is evidence for a muscle mitochondrial defect in lipid oxidation.[25]

Thus, metabolic syndrome is strongly associated with lipid deposition into skeletal muscle, liver, islets, and other organs, suggesting that the accumulation of lipid in subcutaneous adipose tissue depots may lead to excess weight and obesity, but may not necessarily lead to obesity-related metabolic complications unless there is simultaneous lipid deposition into other organs. Since metabolic syndrome is controlled by complex genetics, these data would suggest that the tendency to accumulate lipid in ectopic sites is also under genetic control. Based on the hunter-gatherer lifestyle during early human development, and the survival advantage conferred upon an individual with efficient fat storage, lipid deposition into ectopic sites may have been desirable as an adjunct to adipose tissue storage. Unfortunately, these evolutionary forces were designed for a human population exposed to intermittent starvation in a setting of extreme physical activity, which is a far cry from the lifestyle of modern humans in developed countries.

ADIPOSE TISSUE SECRETORY PRODUCTS AND INFLAMMATION

Adipose tissue is viewed not simply as a passive lipid storage depot but instead as a highly active metabolic tissue, which secretes numerous products that affect insulin resistance either through a traditional (circulating) hormonal effect, or through local effects on the adipocyte. The term "adipokines" has been used to describe the numerous adipocyte secretory proteins, which include TNFα, IL-6, leptin, PAI-1, resistin, and adiponectin.

Beginning with the description of TNFα expression by adipose tissue of obese rodents and humans in the mid 1990s,[26-28] many adipokines with inflammatory or immunologic properties have been described in adipose tissue and have been implicated in the pathogenesis of metabolic syndrome (Table 6.2). For example, plasma levels of IL-6 and adipose secreted TNFα are associated with obesity, and with insulin resistance, independent of obesity.[29-31] There are several presumed mechanisms for the actions of TNFα on insulin resistance, including an inhibition of insulin receptor signaling,[32] along with an autocrine effect on adipose tissue causing a stimulation of lipolysis and elevation of plasma NEFA.[33] In other studies, TNF knock-out mice did not become insulin resistant with diet-induced obesity.[34]

A number of other genes are expressed in adipose tissue of obese rodents and humans and may be important in insulin resistance. Leptin is expressed by adipose

Table 6.2 Proteins Expressed by Adipose Tissue

Tumor necrosis factor-α (TNFα)

Interleukin-6 (IL-6)

Angiotensinogen

Plasminogen activator inhibitor-1 (PAI-1)

Macrophage chemoattractant protein-1 (MCP-1)

Resistin

Adiponectin

Adipsin

Leptin

tissue and increased with obesity.[35] In rodents, there is evidence for leptin-mediated stimulation of lipid oxidation in skeletal muscle, myocardium, and β-cells, and leptin has been suggested as a host homeostatic mechanism designed to prevent metabolic syndrome.[36] Resistin is an 11 kDa protein that is expressed exclusively in adipose cells, secreted into plasma, and was discovered as a downregulated gene in response to thiazolidinediones.[37] In rodent studies, insulin action was improved by anti-resistin antibodies and was impaired by treatment of normal mice with recombinant resistin. In humans, there is evidence for a link between gene polymorphisms in the resistin gene and insulin resistance.[38]

An important adipocyte secretory product is adiponectin, which is secreted by adipocytes at a high level and acts as an antidiabetic, anti-inflammatory, and anti-atherogenic adipokine.[39,40] Adiponectin circulates as complex multimeric forms, and two receptors for adiponectin have been described.[41] When placed on high-fat diets, adiponectin knockout mice developed marked insulin resistance that significantly improved following adiponectin supplementation.[42] In humans, blood levels of adiponectin were decreased under conditions of obesity, insulin resistance, type 2 diabetes, and coronary disease.[43–45] Several studies have demonstrated that low plasma adiponectin was an independent predictor of insulin resistance,[45,46] and plasma adiponectin increased following treatment of diabetes with thiazolidinediones and weight loss.[47–50]

The cell of origin of many adipokines has come under renewed scrutiny with the recent demonstration that adipose tissue contains macrophages, which may be responsible for the expression of many inflammatory components, and which infiltrates adipose tissue during obesity.[51,52] The expression of macrophage markers was highly correlated with obesity in rodents, and the majority of the macrophages resident in adipose tissue were due to recruitment from the bone marrow. Once activated, macrophages secrete a host of cytokines such as TNFα, IL-6, and IL-1,[53] and the adipose tissue resident macrophages were responsible for the expression of most of the tissue TNFα and IL-6. In human studies, there was a correlation between adipose macrophage markers and BMI.[52] The description of adipose tissue macrophages further complicates the interpretation of the role of adipose tissue in insulin resistance. Whereas our hypotheses previously involved an adipose tissue–muscle connection, we now must confront the infiltration of adipose tissue with macrophages, and posit an adipocyte–macrophage–muscle cell hypothesis, with ectopic lipid accumulation and impaired glucose transport and insulin action as the end products. According to this hypothesis (Figure 6.3), progressive obesity leads to adipose tissue macrophage infiltration, with consequent overexpression of inflammatory cytokines and metabolic syndrome.[53]

DIVERSION OF LIPID INTO ADIPOSE TISSUE AS A MECHANISM FOR IMPROVEMENT IN LIPOTOXICITY

As described above, the excess accumulation of lipid in muscle, islets, and liver results in many of the features of metabolic syndrome and obesity. Hence, treatments aimed at reducing ectopic lipid accumulation would be expected to improve lipotoxicity and, hence, result in a reduction in some of the features of metabolic syndrome. Fortunately, this improved lipotoxicity has been demonstrated in a number of instances. In humans, weight loss and exercise results in improved insulin sensitivity, a reduction in hepatic glucose production, and improved β-cell response,[54–56] along with improved muscle oxidative capacity.[57] However, since many patients cannot or will not undertake such lifestyle measures, much attention has been devoted to drugs, which may improve diabetes or the metabolic syndrome through improved lipotoxicity.

The thiazolidinedione class of drugs includes pioglitazone and rosiglitazone (and formerly troglitazone). These drugs are used in the treatment of diabetes and result in lower blood glucose levels through improved peripheral insulin sensitivity. The

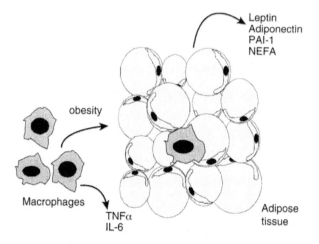

Figure 6.3   Infiltration of adipose tissue with macrophages. With progressive obesity, adipose tissue of some subjects accumulates resident tissue macrophages, which then secrete inflammatory cytokines, and account for much of the inflammatory condition associated with metabolic syndrome.

thiazolidinediones are potent agonists for the transcription factor peroxisome proliferator activated receptor γ (PPARγ), which is found primarily in adipose tissue.
Improved insulin sensitivity is a function that primarily involves glucose transport
into skeletal muscle.[18] Hence, the beneficial effects of these drugs implicate an
adipose tissue–muscle metabolic relationship that is greatly improved by stimulation
of adipocyte PPARγ.

PPARγ activation in adipose tissue would be expected to result in an activation
of adipocyte differentiation, as has been demonstrated *in vitro*.[58,59] Thus, humans
treated with pioglitazone would be expected to demonstrate an increase in adipocyte
differentiation,[60] which would lead to a potential increase in adipose tissue lipid.
The mechanism of this increase in differentiation may be a direct effect of stimulation
of adipogenesis from preadipocytes, or it may be due to a thiazolidinedione-induced
decrease in inflammatory cytokine expression by macrophages. As a result of a
thiazolidinedione-mediated increase in lipid accumulation in adipose tissue, there
may be a diversion of dietary lipid into adipose tissue and away from other organs
that are more susceptible to lipotoxicity, such as muscle, islets, and liver. Thus, the
increase in adipose tissue lipid may, in effect, "steal" lipid from other organs, such
as islets and muscle, which are crucial components of glucose homeostasis.

Several recent studies have provided evidence in favor of this lipid diversion.
In patients with type 2 diabetes, rosiglitazone, but not metformin, resulted in a
reduction in liver fat content.[61] In another study, subjects with impaired glucose
tolerance were treated with either metformin or pioglitazone, followed by muscle
biopsies and the assessment of intramyocellular lipid. There was no change in
intramyocellular lipid with metformin, but there was a significant decrease in
intramyocellular lipid with pioglitazone treatment, which accompanied an
improvement in insulin sensitivity.[62] This reduction in intramyocellular lipid was
accompanied by an increase in the ration of subcutaneous/visceral adipose tissue
mass, as estimated using CT scans, and no evidence for an increase in muscle
lipid oxidation. Thus, these data suggest that the thiazolidinediones may promote
adipocyte differentiation or lipid accumulation and, hence, divert lipid into adipose
tissue and away from muscle, resulting in less lipotoxicity.

ADIPOSE TISSUE: FRIEND OR FOE?

Most modern humans regard the bulges of adipose tissue on their hips, waists, and
thighs as cosmetically unacceptable and a threat to their health. Although the precise
etiology of metabolic disturbance resulting from obesity is unknown, it may stem
from an accident of our evolution and the rapid changes that have occurred in our
lifestyle. Modern humans still have the genome of a hunter-gatherer, trapped in a
body designed for a struggle against famine, infectious diseases, and the threats from
the elements. These Paleolithic threats seldom emerge, and the new threat is the
result of a hunter-gatherer genome faced with an overabundant food supply and no
need for physical activity.

It is likely that the "master planner" of the human genome never imagined a
BMI over 25 kg/m^2, except for perhaps brief periods of time when the hunting was

good and the harvest was truly abundant. In the environment surrounding most human evolution, where the human BMI was always low, physical activity was constant, and starvation was a threat, the ability to store lipid in non-adipose organs may have represented a survival adaptation. Liver and muscle lipid may have represented one more depot for energy storage in anticipation of the next famine. The accumulation of macrophages in adipose tissue may have represented another line of immunologic defense against injury and disease. All of these adaptations developed in a backdrop of leanness and high physical activity, where there was no possibility of obesity, and where the degree of ectopic fat did not reach extreme levels — or only did so with advanced age and post-reproduction years.

Most modern humans are not desirous of returning to the lifestyle of the hunter-gatherer, and unfortunately we are stuck with our hunter-gatherer genome in an easy-living fast food environment. Although it is certainly important that public policies be devoted toward an improvement in lifestyle factors, new drug development should concentrate on the ability of adipose tissue expansion to divert lipid from more toxic sites, coupled with treatments aimed at increasing lipid oxidation.

Fat need not be our enemy. It simply needs to know its place.

ACKNOWLEDGMENTS

This research is supported by a Merit Review Grant from the Veterans Administration, DK 39176 from the National Institutes of Health, and grant number M01RR14288 of the General Clinical Research Center.

REFERENCES

1. Finkelstein, E.A., Fiebelkorn, I.C., and Wang, G. 2003. National medical spending attributable to overweight and obesity: How much, and who's paying? *Health Affairs.* W3:219–226.
2. Mokdad, A.H., Ford, E.S., Bowman, B.A., Dietz, W.H., Vinicor, F., Bales, V.S., and Marks,J.S. 2003. Prevalence of obesity, diabetes, and obesity-related health risk factors, 2001. *JAMA.* 289:76–79.
3. Friedman, J. 2002. Fat in all the wrong places. *Nature.* 415:268–269.
4. Lew, E.A. and Garfinkel, L. 1979. Variations in mortality by weight among 750,000 men and women. *J.Chronic Dis.* 32:563–583.
5. Colditz, G.A., Willett, W.C., Stampfer, M.J., Manson, J.E., Hennekens, C.H., Arky, R.A., and Speizer, F.E. 1990. Weight as a risk factor for clinical diabetes in women. *Am. J. Epidemiol.* 132:501–513.
6. Harris, M.I., Flegal, K.M., Cowie, C.C., Eberhardt, M.S., Goldstein, D.E., Little, R.R., Wiedmeyer, H.M., and Byrd-Holt, D.D. 1998. Prevalence of diabetes, impaired fasting glucose, and impaired glucose tolerance in U.S. adults. The Third National Health and Nutrition Examination Survey, 1988–1994. *Diabetes Care.* 21:518–524.
7. Harris, M.I. 1998. Diabetes in America: epidemiology and scope of the problem. *Diabetes Care.* 21:Suppl-4.

8. Kissebah, A.H., Freedman, D.S., and Peiris, A.N. 1989. Health risks of obesity. *Med. Clin. North Am.* 73:111–138.
9. Bray, G.A. 2004. Medical Consequences of Obesity. *J. Clin. Endocrinol. Metab.* 89:2583–2589.
10. Kissebah, A.H. 1991. Insulin resistance in visceral obesity. *Int. J. Obes.* 15 Suppl. 2:109–115.
11. Klein, S., Fontana, L., Young, V.L., Coggan, A.R., Kilo, C., Patterson, B.W., and Mohammed, B.S. 2004. Absence of an effect of liposuction on insulin action and risk factors for coronary heart disease. *N. Engl. J. Med.* 350:2549–2557.
12. Klein, S. 2004. The case of visceral fat: argument for the defense. *J. Clin. Invest.* 113:1530–1532.
13. DeFronzo, R.A., Bonadonna, R.C., and Ferrannini, E. 1992. Pathogenesis of NIDDM. A balanced overview. *Diabetes Care.* 15:318–368.
14. Bergman, R. 1989. Toward physiological understanding of glucose tolerance: minimal model approach. *Diabetes.* 39:1512–1527.
15. Kahn, S.E., Prigeon, R.L., McCulloch, D.K., Boyko, E.J., Bergman, R.N., Schwartz, M.W., Neifing, J.L., Ward, W.K., Beard, J.C., and Palmer, J.P. 1993. Quantification of the relationship between insulin sensitivity and beta-cell function in human subjects. Evidence for a hyperbolic function. *Diabetes.* 42:1663–1672.
16. Gerich, J.E. 1998. The genetic basis of type 2 diabetes mellitus: impaired insulin secretion versus impaired insulin sensitivity. *Endocr. Rev.* 19:491–503.
17. Lee, Y., Hirose, H., Ohneda, M., Johnson, J.H., McGarry, J.D., and Unger, R.H. 1994. Beta-cell lipotoxicity in the pathogenesis of non-insulin-dependent diabetes mellitus of obese rats: impairment in adipocyte-beta-cell relationships. *Proc. Natl. Acad. Sci. USA.* 91:10878–10882.
18. DeFronzo, R.A., Jacot, E., Jequier, E., Maeder, E., Wahren, J., and Felber, J.P. 1981. The effect of insulin on the disposal of intravenous glucose. Results from indirect calorimetry and hepatic and femoral venous catheterization. *Diabetes.* 30:1000–1007.
19. Lithell, H., Lindgarde, F., Hellsing, K., Lundquist, G., Nygarrd, E., Vessby, B., and Saltin, B. 1981. Body weight, skeletal muscle morphology and enzyme activities in relation to fasting serum insulin concentration and glucose tolerance in 48 year old men. *Diabetes.* 30:19–25.
20. Lillioja, S., Young, A.A., Culter, C.L., Ivy, J.L., Abbott, W.G.H., Zawadzki, J.K., Yki-Jarvinen, H., Christin, L., Secomb, T.W., and Bogardus, C. 1987. Skeletal muscle capillary density and fiber type are possible determinants of in vivo insulin resistance in man. *J. Clin. Invest.* 80:415–424.
21. McGarry, J.D. 1994. Disordered metabolism in diabetes: have we underemphasized the fat component? *J. Cell. Biochem.* 55:Suppl:29–38.
22. Unger, R.H. 1995. Lipotoxicity in the pathogenesis of obesity-dependent NIDDM. Genetic and clinical implications. *Diabetes.* 44:863–870.
23. Simoneau, J.A., Colberg, S.R., Thaete, F.L., and Kelley, D.E. 1995. Skeletal muscle glycolytic and oxidative enzyme capacities are determinants of insulin sensitivity and muscle composition in obese women. *FASEB Journal.* 9:273–278.
24. Falholt, K., Jensen, I., Lindkaer, J.S., Mortensen, H., Volund, A., Heding, L.G., Noerskov, P.P., and Falholt, W. 1988. Carbohydrate and lipid metabolism of skeletal muscle in type 2 diabetic patients. *Diabetic Med.* 5:27–31.
25. Petersen, K.F., Dufour, S., Befroy, D., Garcia, R., and Shulman, G.I. 2004. Impaired mitochondrial activity in the insulin-resistant offspring of patients with type 2 diabetes. *N. Engl. J. Med.* 350:664–671.

26. Hotamisligil, G.S., Shargill, N.S., and Spiegelman, B.M. 1993. Adipose expression of tumor necrosis factor-α: Direct role in obesity-linked insulin resistance. *Science.* 259:87–91.

27. Hotamisligil, G.S., Arner, P., Caro, J.F., Atkinson, R.L., and Spiegelman, B.M. 1995. Increased adipose tissue expression of tumor necrosis factor-alpha in human obesity and insulin resistance. *J. Clin. Invest.* 95:2409–2415.

28. Kern, P.A., Saghizadeh, M., Ong, J.M., Bosch, R.J., Deem, R., and Simsolo, R.B. 1995. The expression of tumor necrosis factor in human adipose tissue. Regulation by obesity, weight loss, and relationship to lipoprotein lipase. *J. Clin. Invest.* 95:2111–2119.

29. Kern, P.A., Ranganathan, S., Li, C., Wood, L., and Ranganathan, G. 2001. Adipose tissue tumor necrosis factor and interleukin-6 expression in human obesity-associated insulin resistance. *Am. J. Physiol. Endocrinol. Metab.* 280:E745–E751.

30. Mohamed-Ali, V., Goodrick, S., Rawesh, A., Katz, D.R., Miles, J.M., Yudkin, J.S., Klein, S., and Coppack, S.W. 1997. Subcutaneous adipose tissue releases interleukin-6, but not tumor necrosis factor-a, in vivo. *J. Clin. Endocrinol. Metab.* 82:4196–4200.

31. Bastard, J.P., Jardel, C., Bruckert, E., Blondy, P., Capeau, J., Laville, M., Vidal, H., and Hainque, B. 2000. Elevated levels of interleukin 6 are reduced in serum and subcutaneous adipose tissue of obese women after weight loss. *J. Clin. Endocrinol. Metab.* 85:3338–3342.

32. Hotamisligil, G.S., Murray, D.L., Choy, L.N., and Spiegelman, B.M. 1994. Tumor necrosis factor inhibits signaling from the insulin receptor. *Proc. Natl. Acad. Sci. USA.* 91:4854–4858.

33. Patton, J.S., Shepard, H.M., Wilking, H., Lewis, G., Aggarwal, B.B., Eessalu, T.E., Gavin, L.A., and Grunfeld, C. 1986. Interferons and tumor necrosis factors have similar catabolic effects on 3T3-L1 cells. *Proc. Natl. Acad. Sci. USA.* 83:8313–8317.

34. Uysal, K.T., Wiesbrock, S.M., Marino, M.W., and Hotamisligil, G.S. 1997. Protection from obesity-induced insulin resistance in mice lacking TNF-alpha function. *Nature.* 389:610–614.

35. Maffei, M., Halaas, J., Ravussin, E., Pratley, R.E., Lee, G.H., Zhang, Y., Fei, H., Kim, S., Lallone, R., Ranganathan, S. et al. 1995. Leptin levels in human and rodent: measurement of plasma leptin and ob RNA in obese and weight-reduced subjects. *Nature Med.* 1:1155–1161.

36. Unger, R.H. 2002. Lipotoxic diseases. *Ann. Rev. Medicine.* 53:319–336.

37. Steppan, C.M., Bailey, S.T., Bhat, S., Brown, E.J., Banerjee, R.R., Wright, C.M., Patel, H.R., Ahima, R.S., and Lazar, M.A. 2001. The hormone resistin links obesity to diabetes. *Nature.* 409:307–312.

38. Wang, H., Chu, W.S., Hemphill, C., and Elbein, S.C. 2002. Human resistin gene: molecular scanning and evaluation of association with insulin sensitivity and type 2 diabetes in Caucasians. *J. Clin. Endocrinol. Metab.* 87:2520–2524.

39. Scherer, P.E., Williams, S., Fogliano, M., Baldini, G., and Lodish, H.F. 1995. A novel serum protein similar to C1q, produced exclusively in adipocytes. *J. Biol. Chem.* 270:26746–26749.

40. Kumada, M., Kihara, S., Sumitsuji, S., Kawamoto, T., Matsumoto, S., Ouchi, N., Arita, Y., Okamoto, Y., Shimomura, I., Hiraoka, H. et al. 2003. Association of hypo-adiponectinemia with coronary artery disease in men. *Arterioscler. Thromb. Vasc. Biol.* 23:85–89.

41. Pajvani, U.B., Du, X., Combs, T.P., Berg, A.H., Rajala, M.W., Schulthess, T., Engel, J., Brownlee, M., and Scherer, P.E. 2003. Structure-function studies of the adipocyte-secreted hormone Acrp30/adiponectin. Implications for metabolic regulation and bioactivity. *J. Biol. Chem.* 278:9073–9085.

42. Maeda, N., Shimomura, I., Kishida, K., Nishizawa, H., Matsuda, M., Nagaretani, H., Furuyama, N., Kondo, H., Takahashi, M., Arita, Y. et al. 2002. Diet-induced insulin resistance in mice lacking adiponectin/ACRP30. *Nature Med.* 8:731–737.

43. Hotta, K., Funahashi, T., Arita, Y., Takahashi, M., Matsuda, M., Okamoto, Y., Iwahashi, H., Kuriyama, H., Ouchi, N., Maeda, K. et al. 2000. Plasma concentrations of a novel, adipose-specific protein, adiponectin, in type 2 diabetic patients. *Arterioscler. Thromb. Vasc. Biol.* 20:1595–1599.

44. Arita, Y., Kihara, S., Ouchi, N., Takahashi, M., Maeda, K., Miyagawa, J., Hotta, K., Shimomura, I., Nakamura, T., Miyaoka, K. et al. 1999. Paradoxical decrease of an adipose-specific protein, adiponectin, in obesity. *Biochem. Biophys. Res. Commun.* 257:79–83.

45. Kern, P.A., Di Gregorio, G.B., Lu, T., Rassouli, N., and Ranganathan, G. 2003. Adiponectin expression from human adipose tissue: relation to obesity, insulin resistance, and tumor necrosis factor-α expression. *Diabetes.* 52:1779.

46. Weyer, C., Funahashi, T., Tanaka, S., Hotta, K., Matsuzawa, Y., Pratley, R.E., and Tataranni, P.A. 2001. Hypoadiponectinemia in obesity and type 2 diabetes: close association with insulin resistance and hyperinsulinemia. *J. Clin. Endocrinol. Metab.* 86:1930–1935.

47. Yang, W.S., Lee, W.J., Funahashi, T., Tanaka, S., Matsuzawa, Y., Chao, C.L., Chen, C.L., Tai, T.Y., and Chuang, L.M. 2001. Weight reduction increases plasma levels of an adipose-derived anti-inflammatory protein, adiponectin. *J. Clin. Endocrinol. Metab.* 86:3815–3819.

48. Phillips, S.A., Ciaraldi, T.P., Kong, A.P.S., Bandukwala, R., Aroda, V., Carter, L., Baxi, S., Mudaliar, S.R., and Henry, R.R. 2003. Modulation of circulating and adipose tissue adiponectin levels by antidiabetic therapy. *Diabetes.* 52:667–674.

49. Yang, W.S., Jeng, C.Y., Wu, T.J., Tanaka, S., Funahashi, T., Matsuzawa, Y., Wang, J.P., Chen, C.L., Tai, T.Y., and Chuang, L.M. 2002. Synthetic peroxisome proliferator-activated receptor-gamma agonist, rosiglitazone, increases plasma levels of adiponectin in type 2 diabetic patients. *Diabetes Care.* 25:376–380.

50. Maeda, N., Takahashi, M., Funahashi, T., Kihara, S., Nishizawa, H., Kishida, K., Nagaretani, H., Matsuda, M., Komuro, R., Ouchi, N. et al. 2001. PPARgamma ligands increase expression and plasma concentrations of adiponectin, an adipose-derived protein. *Diabetes.* 50:2094–2099.

51. Xu, H., Barnes, G.T., Yang, Q., Tan, G., Yang, D., Chou, C.J., Sole, J., Nichols, A., Ross, J.S., Tartaglia, L.A. et al. 2003. Chronic inflammation in fat plays a crucial role in the development of obesity-related insulin resistance. *J. Clin. Invest.* 112:1821–1830.

52. Weisberg, S.P., McCann, D., Desai, M., Rosenbaum, M., Leibel, R.L., and Ferrante, A.W., Jr. 2003. Obesity is associated with macrophage accumulation in adipose tissue. *J. Clin. Invest.* 112:1796–1808.

53. Wellen, K.E. and Hotamisligil, G.S. 2003. Obesity-induced inflammatory changes in adipose tissue. *J. Clin. Invest.* 112:1785–1788.

54. Henry, R.R. and Gumbiner, B. 1991. Benefits and limitations of very-low-calorie diet therapy in obese NIDDM. *Diabetes Care.* 14:802–823.

55. Petersen, K.F., Dufour, S., Befroy, D., Lehrke, M., Hendler, R.E., and Shulman, G.I. 2005. Reversal of nonalcoholic hepatic steatosis, hepatic insulin resistance, and hyperglycemia by moderate weight reduction in patients with type 2 diabetes. *Diabetes.* 54:603–608.
56. Goodpaster, B.H., Katsiaras, A., and Kelley, D.E. 2003. Enhanced fat oxidation through physical activity is associated with improvements in insulin sensitivity in obesity. *Diabetes.* 52:2191–2197.
57. Kern, P.A., Simsolo, R.B., and Fournier, M. 1999. Effect of weight loss on muscle fiber type, fiber size, capillarity, and succinate dehydrogenase activity in humans. *J. Clin. Endocrinol. Metab.* 84:4185–4190.
58. Spiegelman, B.M. 1998. PPAR-γ: adipogenic regulator and thiazolidinedione receptor. *Diabetes.* 47:507–514.
59. Tontonoz, P., Hu, E., and Spiegelman, B.M. 1994. Stimulation of adipogenesis in fibroblasts by PPAR gamma 2, a lipid-activated transcription factor. *Cell.* 79:1147–1156.
60. Okuno, A., Tamemoto, H., Tobe, K., Ueki, K., Mori, Y., Iwamoto, K., Umesono, K., Akanuma, Y., Fujiwara, T., Horikoshi, H. et al. 1998. Troglitazone increases the number of small adipocytes without the change of white adipose tissue mass in obese Zucker rats. *J. Clin. Invest.* 101:1354–1361.
61. Tiikkainen, M., Hakkinen, A.M., Korsheninnikova, E., Nyman, T., Makimattila, S., and Yki-Jarvinen, H. 2004. Effects of rosiglitazone and metformin on liver fat content, hepatic insulin resistance, insulin clearance, and gene expression in adipose tissue in patients with type 2 diabetes. *Diabetes.* 53:2169–2176.
62. Rasouli, N., Raue, U., Miles, L.M., Lu, T., Di Gregorio, G.B., Elbein, S.C., and Kern, P.A. 2005. Pioglitazone improves insulin sensitivity through reduction in muscle lipid and redistribution of lipid into adipose tissue. *Am. J. Physiol. Endocrinol. Metab.* 288:E930–E934.

CHAPTER 7

Gene–Environment Interactions in Obesity: Implications for the Prevention and Treatment of Obesity

Louis Pérusse

CONTENTS

INTRODUCTION

Overweight, defined as a body mass index (BMI: body weight in kg divided by the square of the height in m) between 25 and 30 kg/m², and obesity, defined as a BMI of 30 kg/m² and above, are prevalent conditions in industrialized populations. According to a recent report from the World Health Organization,[1] the prevalence of both conditions has reached epidemic proportions worldwide, with rates of about 50 to 60% of the adult population in the U.S., Canada, and European countries.[2-6] The health consequences of obesity are enormous. Obesity is associated with some of the most prevalent diseases of modern societies, including type 2 diabetes, hypertension, cardiovascular diseases and certain types of cancer (breast, uterus, prostate, colon). The increase in the prevalence of obesity observed worldwide in the past 50 years has occurred in a changing environment characterized by a progressive reduction in energy expenditure associated with physical activity and the abundance of highly palatable foods. These environmental changes occurred over a period of time that is too short to cause changes in the frequencies of genes associated with obesity. Thus, genes that were selected for energy storage in the primitive hunter/gatherer populations are now detrimental in an era of food abundance. From a genetic point of view, this suggests that gene–environment interactions are important in determining an individuals' susceptibility to obesity and related co-morbidities.

This chapter provides an overview of the role of gene–environment interactions in obesity. The evidence supporting the role of genetic factors in obesity will first be reviewed followed by a brief description of the methods used to detect gene–environment interactions in human populations and a review of the evidence for a role of gene–environment interactions in obesity. The implications of gene–environment interactions for the prevention and treatment of obesity will also be discussed.

GENETIC AND NON-GENETIC DETERMINANTS OF OBESITY

Before investigating the issue of gene–environment interactions in obesity, it must first be established that obesity is a condition characterized by familial resemblance and influenced by genetic factors. It is well established that obesity runs in families, suggesting that individuals with a positive family history of obesity are at increased risk of obesity. This risk can be quantified by calculating the ratio of the risk of being obese when a biological relative is obese to the risk in the population at large (i.e., the prevalence of obesity).[7] Studies have shown that risk of obesity is about 2 to 8 times higher in families of obese individuals than in the population at large, a risk that tends to increase with the severity of obesity.[8–11] This familial risk was estimated for various indicators of obesity in the Canadian population using data from 6,377 families (15,245 subjects) who participated in the 1981 Canada Fitness Survey.[10] The indicators of obesity included the body mass index (BMI), the sum of 5 skinfold thickness (SF5), the ratio of the trunk-to-extremity skinfolds ratio (TER), and waist circumference (WC). Figure 7.1 presents the age- and sex-standardized risk ratios (SSRs) comparing the prevalence rates for the various obesity indicators in spouses and first-degree relatives of probands who exceeds the 99th percentile of the age- and sex-specific distributions of the indicators. Except for SF5, the SSRs are higher in first-degree relatives than in spouses, suggesting a greater role of genetic factors. The comparison of the SSRs between spouses and first-degree relatives reveals that the SSRs are reduced to 1.0 for TER and WC, while they remain to a value of about 3.0 for BMI. As spouses do not share genes by descent, this suggests that the contribution of genetic factors is probably more important for indicators of abdominal obesity than for indicators of body mass. These data clearly suggest that obesity is a condition that runs in families. We have estimated from these data that approximately 30 to 40% of the variability in various indicators of obesity and fat distribution could be explained by the transmission of genetic and cultural factors from parents to offspring.[12]

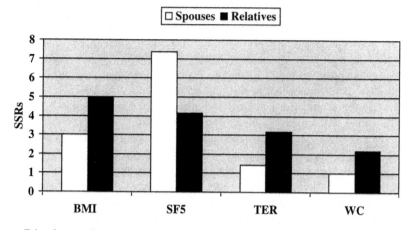

Figure 7.1 Age- and sex-standardized risk ratios comparing the prevalence of various indicators of obesity in spouses and first-degree relatives of probands exceeding the 99th percentile of the distribution. BMI = body mass index; SF5 = sum of 5 skinfolds (biceps + triceps + subscapular + suprailiac + calf); TER = trunk-to-extremity skinfold ratio adjusted for SF5; WC = waist circumference adjusted for BMI. (Adapted from Katzmarzyk et al., *Am. J. Epidemiol.*, 1999, 149:933–942.)

Several studies have been conducted over the past 20 years to determine the relative contribution of genetic vs. familial environmental factors in this familial aggregation. An overview of these studies from several review papers[13–19] indicates that heritability estimates vary from about 0 to 90% depending on the obesity phenotype investigated and the study design and methods used to assess the heritability. In general, heritability estimates are highest (around 70%) when derived from twin studies[20,21] and lowest (around 10%) in adoption studies[16,22] and tend to be higher for phenotypes indexing fat distribution than for phenotypes indexing total body mass or body fat.[23] The results obtained in the Quebec Family Study (QFS) and reviewed by Pérusse at al.,[18] indicate heritability estimates of about 25 to 40% for the amount of body fat (assessed by underwater weighing) and about 40 to 50% for phenotypes indexing fat distribution, like waist circumference or abdominal fat derived from measures of trunk skinfolds. Studies with direct measures of abdominal fat obtained by computed tomography showed that about 50 to 60% of the variance in the amount of abdominal fat could be accounted for by genetic factors.[24,25]

The yearly updates of the obesity gene map published over the past 10 years provide a good indication of the progress accomplished in the past decade in the search for genes and DNA sequence variations associated with human obesity. One of the strongest pieces of evidence for a role of genes in human obesity is the existence of monogenic forms of obesity. These rare and severe forms of obesity are due to deficiencies in single genes. According to the most recent obesity gene map,[26] a total of 173 human obesity cases due to single-gene mutations in 10 different genes have been reported worldwide. The gene that has been the most frequently associated with these monogenic forms of obesity is the melanocortin-4 receptor (MC4R) gene, with 51 mutations accounting for 82% of the monogenic forms of obesity. MC4R is expressed in the brain, in a region of the hypothalamus involved in the regulation of food intake. Another line of evidence supporting the role of genetic factors in obesity comes from studies that have reported positive associations between candidate genes and various obesity-related phenotypes. Based on the literature reviewed up to the end of October 2004, a total of 358 studies reported positive associations with 113 candidate genes.[26] Those that have been most frequently associated with obesity include the adrenergic receptor beta-3 (ADRB3) and the peroxisome proliferator-activated receptor-gamma (PPARG) genes with more than 20 studies with positive findings for each of them.

Despite strong evidence showing that genes could play a role in obesity, the variance accounted for by genetic factors for obesity-related phenotypes rarely exceeds 50%, which suggests that environmental factors are also important in the etiology of obesity. The WHO report on obesity[1] identified physical activity and the availability of high-fat energy-dense foods as the two principal modifiable environmental factors responsible for the increased prevalence of obesity observed worldwide. Results from several cross-sectional studies reviewed by Hill et al.[27] show that there is an inverse relationship between physical activity level and various indicators of obesity and that regular physical activity is protective against obesity. Because of the alarming trends noted in the rates of childhood obesity,[28] the issue of the relationship between physical activity and obesity has received considerable attention. In children, sedentary behaviors are often assessed by the time spent watching television and playing video games. Several studies have investigated the association between television viewing and children's risk of obesity. Generally, hours of television viewing are closely associated with increased levels of obesity both in cross-sectional and prospective studies.[28] For example, a prospective study of 700 children aged 10 to 15 years followed for 4 years showed a strong dose-response relationship between hours of television viewing and prevalence of overweight: children watching television more than 5 hours a day were 5 times more likely to be overweight than those watching fewer than 2 hours a day.[29] Despite these positive associations, it is important to keep in mind that television viewing is not a direct correlate of the overall physical activity level of children. The association between television viewing and body fatness and physical activity level was recently reviewed using a meta-analysis.[30] The association with body fatness was investigated with data from 30 studies including more than 44,000 children, while the association with physical activity was investigated using data from 33 studies including more than 143,000 children. The authors reported a positive association between television viewing and body fatness and a negative association with physical activity level but concluded that the associations were probably too small to be of substantial clinical relevance.[30]

The role of dietary fat in the etiology of obesity has been addressed in several studies but remains controversial. It is generally accepted that high-fat diets induce an overconsumption of energy, which can lead to the development of obesity over time. However, the question as to whether or not a high-fat diet by itself (i.e., independent of total energy intake) is a risk factor for obesity remains controversial. A recent review of the available evidence concluded that despite the lack of a study showing a definitive causal relationship, studies consistently show that high-fat diets increase the likelihood of obesity and that the risk of obesity is low in individuals consuming low-fat diets.[31]

DETECTION OF GENE–ENVIRONMENT INTERACTIONS IN HUMANS

Data reviewed above clearly indicate that genetic and environmental factors, partic-ularly diet and physical activity, contribute to obesity. In addition to the main effects of genes and environment, there may be interactions among these factors. These effects are generally non-additive and not considered in the studies reviewed above. From a statistical point of view, gene–environment interaction (G×E) corresponds to a departure from expectation of the joint effect of the genetic and environmental determinants of a trait or a disease. This means that attempts to estimate the relative contribution of genetic and environmental factors to a disease while ignoring their interactions could lead to incorrect estimates of the proportion of the disease that is explained by genes, the environment, and their joint effects.[32] In the context of obesity, G×E may occur when the response of an obesity phenotype to an intervention (diet or exercise) depends on (or is modified by) an individual's genotype. There are two major approaches used to study G×E in humans — unmeasured genotype and measured genotype, as described previously.[33]

The unmeasured genotype approach is based on statistical analysis of the distri-bution of phenotypes in individuals and families and does not rely on any direct measure of DNA variation. Evidence of G×E using the unmeasured approach can be obtained by comparing the effect of an environmental exposure in individuals with and without a genetic susceptibility to the disease being investigated. In that case, genetic susceptibility is inferred from family history (presence or absence of a positive family history of disease) or a phenotype (skin color, for example). The twin methodology is a very useful unmeasured genotype approach for testing G×E. The effect of an environmental factor on a phenotype can be tested for significant differences between monozygotic (MZ) and dizygotic (DZ) twins, which share on average, 100% and 50% of their genes, respectively. Evidence that the effects of the environment differ across twin types is an indication of a G×E effect. We have proposed that another way to test for the presence of a G×E using the unmeasured genotype approach is to perform an intervention study in monozygotic (MZ) twin pairs, where MZ pairs (who share 100% of their genes in common) are challenged under standardized treatments.[34] A comparison of the within- and between-pair variances of the response to treatment provides an indication of whether genetic factors underlie the response. That is, a greater variability between- than within-pairs suggests a greater correlated response to environmental challenge.

The measured genotype approach uses genetic variation in random genetic mark-ers or in candidate genes and attempts to evaluate the impact of variation at the DNA level on the quantitative phenotype under study. With the advent of methods from molecular biology and the completion of the human genome sequence, gene–environment interactions are now almost exclusively studied using the mea-sured genotype approach. The method is usually applied to association analysis of candidate genes. Several designs can be used to assess gene–environment interaction using the candidate gene approach.[32] In association analysis of candidate genes, the risk of disease is compared among groups stratified by genotype (e.g., carriers vs. non-carriers of a genetic variant increasing the risk of disease) and environmental exposure (exposed vs. non-exposed). In this case, the environmental exposure and genotype are treated as dichotomous variables and gene–environment interaction involves testing whether the relative risk for joint exposure is significantly greater than would be expected by multiplying the relative risks for environmental exposure or genetic susceptibility alone.[32] Environmental exposure can also be treated as a continuous variable and included as a main effect in a regression model, which also incorporates the genotype as well as the interaction term between them. The dependent variable can be either dichotomous (obese vs. non-obese) or continuous (e.g., body mass index). Finally, GxE can be tested using an intervention design by comparing the response to intervention among groups stratified by genotypes. If individuals with a particular genotype at candidate gene tend to respond to the intervention differently than do individuals with alternative genotypes, evidence of gene–environment interaction can be inferred.

EVIDENCE OF GENE–ENVIRONMENT INTERACTION IN OBESITY

Evidence from the Unmeasured Genotype Approach

One longitudinal study of obesity, dietary fat, and familial predisposition to obesity conducted in women provided indirect evidence of gene–diet interaction for obesity.[35] After controlling for total energy intake and other factors, such as smoking, activity, menopausal status, a high dietary fat intake was associated with 6-year changes in BMI, but only among women with a familial history of obesity. In another longitudinal study, the Finnish Twin Cohort Study was used to examine whether physical activity was playing a role in 6-year body weight changes in 1,571 MZ and 3,029 DZ same-sex twin pairs.[36] The results showed that associations between weight change in twin A and twin B were significantly stronger for monozygotic than for same-sex dizygotic twins at all levels of physical activity, suggesting an interaction between genotype and physical activity level for changes in body weight.

More direct evidence of GxE on obesity was obtained from intervention studies conducted in MZ twins. We have performed a series of studies to investigate the role of the genotype in determining the response to changes in energy balance by submitting both members of male MZ twin pairs either to positive energy balance induced by overfeeding[37] or negative energy balance induced by exercise training[38] in presence of clamped energy intake conditions. In the overfeeding experiment, 12 pairs of healthy male MZ twins were submitted to a 1000 kcal caloric surplus, 6 days a week, for a period of 100 days for an excess energy intake over the entire protocol of 84,000 kcal. In the negative energy balance experiment, 7 pairs of male MZ twins exercised on cycle ergometers twice a day, 9 out of 10 days, over a period of 93 days while being kept on a constant daily energy and macro nutrient intake. The mean total energy deficit caused by exercise above the estimated energy cost of body weight maintenance reached about 58,000 kcal. The results of these experiments are summarized in Table 7.1. Mean body gain was 8.1 kg with a range of about 4 to 13 kg, while mean body weight loss was about 5.0 kg with a range of about 1 kg to about 8 kg. As indicated by the F ratios, the variance in the response was about 3 to 14 times higher between pairs than within pairs depending on the

Table 7.1 Intrapair Resemblance in the Response of Obesity Phenotypes to Overfeeding and Negative Energy Balance Protocols in Monozygotic Twins

Phenotype	Overfeeding				Negative Energy Balance			
	Changes	Min / Max	Within-pair resemblance		Changes	Min / Max	Within-pair resemblance	
			F Ratio	R intra			F Ratio	R intra
Body weight (kg)	8.1 ± 2.3	4.3 / 13.3	3.4*	0.55	-4.9 ± 2.2	-8.1 / -1.1	6.8*	0.74
BMI (kg/m²)	2.6 ± 0.7	1.2 / 4.2	2.8*	0.48	-1.6 ± 0.6	-2.4 / -0.3	6.1*	0.72
Fat mass (kg)	5.4 ± 1.9	1.4 / 9.6	3.0*	0.50	-4.9 ± 2.3	-8.8 / -1.3	14.1**	0.87
Percent body fat	6.5 ± 2.4	2.0 / 12.3	2.9*	0.49	-4.8 ± 2.1	-7.6 / -1.5	9.0**	0.80
AVF (cm²)	24.4 ± 12.6	-0.3 / 45.2	6.1**	0.72	-28.7 ± 13.0	-50.9 / -11.4	11.7**	0.84

Values are means ± SD. BMI = body mass index. AVF = abdominal visceral fat from CT scan.

The changes observed in all phenotypes following overfeeding and negative energy balance were significant ($p < 0.01$).

* $p < 0.05$; ** $p < 0.01$

Source: Adapted from Bouchard, C. et al., *N. Engl. J. Med.* 1990, 322:1477–1482 and Bouchard, C. et al., *Obes. Rev.* 1994, 2:400–410.

phenotype and the protocol. The intraclass correlation coefficients used to assess the within-pair resemblance were all significant, ranging from 0.48 to 0.87. These results suggest that the response to a caloric surplus or a caloric deficit is strongly dependent of the individual's genotype. As of today, these results remain the best experimental evidence supporting the role of gene–environment interaction in obesity.

Evidence from the Measured Genotype Approach

Gene–Diet Interactions

Despite the obvious connection between food intake and obesity, there is a paucity of gene–diet interaction studies for phenotypes of obesity. Some results from intervention studies have provided evidence for a role of a few candidate genes in modulating the changes in body fatness following a dietary intervention. Some of this evidence is coming from the overfeeding studies that were conducted in MZ twins and is reviewed in detail elsewhere.[39] A total of 58 polymorphisms in 40 candidate genes were tested for association with changes in various body fatness indicators in response to overfeeding. Table 7.2 provides a summary of the genes that showed significant evidence of association for changes in body weight, fat mass, subcutaneous fat (sum of 8 skinfolds) and abdominal visceral fat. The gene encoding adipsin, a protein secreted by the adipoctyes and found to be elevated in obesity, showed significant association with changes in body weight and body fat accounting for up to 20% of the variance in fat mass changes. The beta-2 adrenergic receptor (ADRB2) gene was found to be strongly associated ($p < 0.005$) with body fatness accounting for about 7% of the gains in body weight and subcutaneous fat. ADRB2 plays an important role in the regulation of energy balance by controlling lipid mobilization in the adipose tissue. The Bcl I variant of the glucocorticoid receptor gene (GRL) was also associated with the overfeeding response accounting for 5% and 8% of the variance in body weight ($p < 0.005$) and abdominal visceral fat ($p < 0.05$) changes, respectively. GRL plays an important role in the regulation of energy balance as it binds cortisol, a hormone inhibiting food intake. A polymorphism in the insulin-like growth factor binding protein-1 (IGFBP1) was also found to be associated with changes in the amount of abdominal visceral fat in response to overfeeding. The role of the resistin gene in determining the response to overfeeding was also examined.[40] Resistin is a hormone secreted exclusively by the adipocytes and involved in the development of insulin resistance associated with obesity. An intronic variant in the resistin gene was found to be associated with changes in total abdominal ($p = 0.03$) and visceral ($p = 0.004$) fat in response to overfeeding.

Few other intervention studies involving candidate genes of obesity support the role of GxE in obesity. Fumeron et al.[41] submitted 163 obese patients to a low-calorie diet and found that a polymorphism in the uncoupling protein-1 (UCP1) gene was associated with a resistance to lose weight, weight loss being lower in patients carrying the less frequent allele (4.5 kg) compared to patients carrying the more frequent allele (7.5 kg). In 211 obese patients undergoing treatment with sibutramine (a drug prescribed to lose weight), weight losses were significantly ($p = 0.001$)

Table 7.2 Candidate Genes Associated with Changes in Body Fat in Response to Overfeeding in Twins

Gene	Chromosome	Body Weight		Fat Mass		Subcutaneous Fat		Abdominal Visceral Fat	
		p value	R^2 (%)	p value	R^2 (%)	p value	R^2 (%)	p value	R^2 (%)
Adipsin	19p13.3	< 0.05	0.6	< 0.05	20.9	< 0.05	1.9	< 0.05	2.4
ADRB2	5q31-q32	< 0.001	7.0	NS		< 0.005	7.1	NS	
GRL	5q31	< 0.005	4.6	NS		NS		< 0.05	7.9
IGF2	11p15.5	NS		NS		< 0.05	2.1	NS	
IGFBP1	7p13	NS		NS		NS		NS	
LPL	8p21.3	< 0.05	5.4	NS		NS		< 0.01	21.5

R^2 = percentage of variance in the changes accounted for by the candidate gene.

Source: Adapted from Ukkola, O. and Bouchard, C., Obes. Rev. 2004, 5:3–12.

associated with the C825T polymorphism of the guanine nucleotide binding protein-3 (GNB3) gene.[42]

Evidence of gene–diet interaction for obesity can also be obtained without intervention using a cross-sectional design. For example, interactions between dietary fat intake and polymorphisms in 11 different candidate genes of obesity were tested on 154 obese subjects and 154 age- and sex-matched normal weight controls.[43] Effects of interactions between candidate genes and dietary fat on the risk of obesity were assessed by means of a logistic regression after adjustment for total energy intake and physical activity. Significant evidence of interaction was found between the Pro12Ala polymorphism of the PPARG gene and dietary intake of arachidonic acid. Individuals who were carriers of the PPARG Ala variant and who consumed high amounts of arachidonic acid had a significantly higher risk of obesity than the Pro12Pro individuals. Significant evidence of interaction with linoleic acid intake was also found with polymorphisms in the leptin (LEP) and the tumor necrosis factor-alpha (TNFA) genes.[43] Results from this study clearly suggest that genetic variation in the LEP, PPARG, and TNFA genes can influence diet-related risk of obesity. We also found evidence of gene–diet interaction with the Pro12Ala polymorphism of the PPARG gene and obesity-related phenotypes in the Quebec Family Study.[44] Total and saturated fat intakes were measured using a 3-day dietary record in a cohort of 720 adult subjects, and interaction with the PPARG Pro12Ala polymorphism was tested for various obesity-related phenotypes using regression procedure. Significant interactions with dietary fat intake were found for BMI (p = 0.001) and waist circumference (p = 0.002). As shown in Figure 7.2, an increase in fat intake was associated with increases in waist circumference, but only in Pro12Pro individuals.

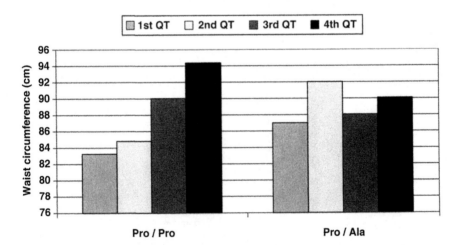

Figure 7.2 Effects of dietary fat intake and PPARG Pro12Ala polymorphism on waist circumference. Quartiles (QT) of total fat intake (g) are: 1 = 22.5 – 65.5; 2 = 65.5 – 86.8; 3 = 86.8 – 100.9; 4 = ≥ 100.9.

Gene–Physical Activity Interactions

A certain number of candidate genes of obesity have been investigated for their role in modulating the response to exercise training or for their interactive effects with exercise or physical activity. Details of these studies can be found in our series of the Human Gene Map for performance and health-related fitness phenotypes reviews published since 2001.[45-48] Table 7.3 presents a summary of the genes involved in interactive effects with physical activity and the response to exercise training. The Gln27Glu polymorphism in the ADRB2 gene was found to be associated with higher body weight, BMI, waist and hip circumferences, and WHR, but only in inactive men and not in those who were physically active. The risk of obesity associated with the Gln27Glu polymorphism was 3.45 (p = 0.002) in sedentary men compared to 1.6 (p = 0.30) in active men.[49] The same polymorphism of the ADRB2 gene was investigated for its effect on the risk of obesity in 139 obese women and 113 normal-weight controls.[50] The effect of the polymorphism on the risk of obesity was tested using a logistic regression procedure taking into account interaction with leisure-time physical activity. A significant (p = 0.005) interaction between physical activity and ADRB2 polymorphism was observed; women who were active in their leisure time and were carriers of the Glu variant had a higher BMI compared to non-carriers, suggesting that the former may be more resistant to exercise-induced weight loss.

Table 7.3   Summary of Candidate Genes Involved in Interactive Effects with Physical Activity and Response to Exercise

Gene	Chromosome	Phenotypes	Reference
		Interaction with Physical Activity	
ADRB2	5q31-q32	Body weight, BMI, waist circumference, waist-to-hip ratio	Meirhaeghe et al.[49]
		Obesity, BMI	Corbalan[50]
UCP3	11q13	BMI	Otabe et al.[51]
		Response to Exercise	
PPARG	3p25	Body weight	Lindi et al.[56]
ADRB2	5q31-q32	BMI, fat mass, percent body fat, subcutaneous fat	Garenc et al.[52]
		Trunk fat, percent body fat	Phares et al.[53]
ADRB3	8p12-p11.2	Fat mass, percent body fat, trunk fat	Phares et al.[53]
		Body weight, BMI, waist-to-hip ratio	Sakane et al.[54]
		Body weight, BMI, waist circumference	Shiwaku et al.[55]
LPL	8p22	BMI, fat mass, percent body fat, abdominal visceral fat	Garenc et al.[57]
CYP19	10q26	BMI, fat mass, percent body fat	Tworoger et al.[58]
UCP3	11q13	Subcutaneous fat	Lanouette et al.[59]
GNB3	12p13	Fat mass, percent body fat	Rankinen et al.[60]
PNMT	17q21-q22	Body weight	Peters et al.[61]
ACE	17q23	Body weight, fat mass	Montgomery et al.[62]
COMT	22q11.21	Percent body fat	Tworoger et al.[58]

Evidence of gene–physical activity interaction was also reported with the UCP3 gene, which was found to modulate the association between BMI and physical activity.[51]

Most of the evidence suggesting a gene–physical activity interaction for obesity comes from intervention studies involving exercise training. Four studies involved polymorphisms in the ADRB2 and ADRB3 genes. In the first study, the effects of the ADRB2 gene on various adiposity phenotypes in response to a 20-week endurance training program were investigated, and results revealed that endurance training resulted in greater reductions of BMI (p = 0.04), fat mass (p = 0.0008) and percent body fat (p = 0.0003), and subcutaneous fat (p = 0.03) in subjects with the Arg16Arg genotype compared to those with the Gly16Gly genotype.[52] In the second study, changes in percent body fat, fat mass, and trunk fat in response to 24 weeks of endurance exercise were tested for associations with polymorphisms in the ADRB2 and ADRB3 genes in 29 sedentary men and 41 postmenopausal women.[53] The Gln27Glu polymorphism of the ADRB2 gene was associated with changes in percent body fat and trunk fat, while the Trp64Arg polymorphism of the ADRB3 gene was associated with changes in fat mass, percent body fat, and trunk fat. In a third study, the effects of the Trp64Arg mutation in the ADRB3 gene on changes in body mass and indicators of body fat distribution following 3 months of a low-calorie and exercise program were investigated in 61 obese women with type 2 diabetes mellitus.[54] Compared to non-carriers, carriers of the mutation were found to have smaller reductions in body weight, BMI, and waist-to-hip ratio. Finally, in a fourth study, the impact of the Trp64Arg polymorphism of the ABDRB3 gene on obesity-related phenotypes in response to a 3-month behavioral intervention using a combination of diet and exercise programs was investigated in 76 middle-aged women.[55] The intervention resulted in a significant reduction of body weight, BMI, and waist circumference, but there were significant differences in the response to the intervention between carriers and non-carriers of the mutant Arg64 allele. Changes in body weight (p = 0.001), BMI (p = 0.002), and waist circumference (p = 0.02) were significant only in wild-type (Trp64Trp) women, which led the authors to conclude that the Trp64Arg mutation of the ADRB3 gene is associated with difficulty in losing weight through behavioral intervention.[55]

A few other genes have been found to modulate the response of body fatness to exercise. The impact of the PPARG Pro12Ala polymorphism on body weight changes was investigated in cohort of 552 obese subjects with impaired glucose tolerance who were randomized to either an intensive diet and exercise group or a control group and followed for a 3-year period.[56] Subjects from the intervention group received individually tailored dietary counseling and advice on physical activity, while subjects from the control group received general information on the benefits of a healthy diet and regular physical activity. In the intervention group, subjects with the Ala12Ala genotype lost more weight during the follow-up (8.3 kg) than subjects with the other genotypes (4 kg). The S447X polymorphism in the LPL showed associations with changes in body fat in response to 20 weeks of endurance training, with greater reductions of body mass index, fat mass, percent body fat, and abdominal visceral fat in women carrying the X447 allele.[57] Another study examined whether polymorphisms in the CYP19 and COMT genes, which encode enzymes regulating the concentrations of estrogen and androgen, were associated with changes in body fatness following a year-long exercise intervention trial in 173 postmenopausal women.[58] One polymorphism in the CYP19 gene was associated with greater reductions in BMI, percent body fat, and fat mass, while a polymorphism in the COMT gene was associated with a smaller decrease in percent body fat. A polymorphism in the UCP3 gene was found to be associated with changes in subcutaneous fat (assessed by the sum of 8 skinfolds in 503 subjects) in response to 20 weeks of endurance training.[59] The C825T polymorphism of the G protein beta-3 (GNB3) was tested for association with changes in body composition following 20 weeks of endurance training,[60] and the TT genotype was found to be associated with a greater decrease in fat mass ($p = 0.012$) and percent body fat ($p = 0.006$). The effect of a polymorphism in the phenylethanolamine N-methyltransferase (PNMT) gene on weight loss was investigated in 149 obese women who participated in a 6-month weight loss trial that included daily intake of a dose of 15 mg of sibutramine and a monthly 1-hour behavior modification seminar that encouraged participants to eat low-fat food, increase their consumption of vegetables and fruits, and to exercise daily.[61] The PNMT gene encodes the rate-limiting enzyme of the conversion of norepinephrine to epinephrine and is thus considered as a candidate gene of sibutramine-induced weight loss, because this drug acts as an inhibitor of the reuptake of norepinephrine in the neurons. A mutation in the promoter of the PNMT gene was associated with greater weight loss in the homozygotes compared to the heterozygotes ($p < 0.002$). Finally, changes in body composition in response to 10 weeks of exercise training were also investigated in 81 subjects as a function of the I/D polymorphism in the angiotensin converting enzyme (ACE) gene.[62] Results indicated that subjects with the II genotype exhibited greater changes in body weight ($p = 0.001$) and fat mass ($p = 0.04$) than carriers of the D allele.

IMPLICATIONS FOR THE PREVENTION AND TREATMENT OF OBESITY

The identification of genes playing a role in interactive effects with environmental factors leading to the development of obesity has several implications for the prevention and treatment of obesity. Some of these implications have been recently summarized[32] and will be briefly discussed in the context of obesity. For the geneticist interested in estimating the contribution of genetic factors in obesity, ignoring interactions can lead to false estimates of the respective contributions of genetic vs. non-genetic determinants of obesity. For the epidemiologist interested in assessing the risk of obesity relative to exposure to an environmental risk factor of obesity (e.g., high-fat diet), ignoring interactions can also lead to false estimates of the risk in exposed vs. non-exposed individuals. For example, if the risk of obesity associated with a high-fat diet is present only in a subgroup of subjects who are genetically susceptible, analysis of the effect of fat intake in the whole group can lead to the false conclusion of a lack of association between dietary fat intake and obesity. Failure to account for gene–environment interactions in obesity might explain the difficulty to replicate the positive findings observed in many candidate gene studies of obesity. The study of gene–environment interaction effects in obesity has the potential to lead to better estimates of the population-attributable risk for genetic and environmental risk factors. The study of gene–environment interactions in obesity also has the potential to increase our understanding of the etiology of obesity by providing information on new biological pathways relevant to obesity and environmental factors most likely to influence these pathways. This information can be used to develop new preventive and therapeutic strategies and eventually to offer patients personalized preventive advice or treatment. It seems likely that, in the near future, it will be possible to use DNA-based tests to determine whether an individual carries genes that increase the susceptibility to gain weight when exposed to a particular diet or increase the resistance to lose weight in response to diet or exercise. The information provided by these DNA tests could then be used to better prevent and treat obesity in individuals having an inherited susceptibility or resistance by altering the appropriate environmental exposure.

REFRENCES

1. WHO. Obesity: preventing and managing the global epidemic. Geneva: World Health Organization, 1998.
2. Laurier D, Guiget M, Chau NP, Wells JA, Valleron AJ. Prevalence of obesity: a comparative survey in France, the United Kingdom and the United States. *Int. J. Obes.* 1992; 16:565–572.
3. Kuskowska-Wolk A, Bergstrom R. Trends in body mass index and prevalence of obesity in Swedish men 1980–89. *J. Epidemiol. Commun. Health.* 1993; 47:103–108.
4. Ogden CL, Troiano RP, Briefel RR, Kuczmarski RJ, Flegal KM, Johnson CL. Prevalence of overweight among preschool children in the United States, 1971 through 1994. *Pediatrics* 1997; 99:1–7.
5. Hanley A, Harris S, Gittelsohn J, Wolever T, Saksvig B, Zinman B. Overweight among children and adolescents in a Native Canadian community: prevalence and associated factors. *Am. J. Clin. Nutr.* 2000; 71:693–700.
6. Kuczmarski RJ. Prevalence of overweight and weight gain in the United States. *Am. J. Clin. Nutr.* 1992; 55:495S–502S.
7. Risch N. Linkage strategies for genetically complex traits. I. Multilocus models. *Am. J. Hum. Genet.* 1990; 46:222–228.
8. Allison DB, Faith MS, Nathan JS. Risch's lambda values for human obesity. *Int. J. Obes.* 1996; 20:990–999.
9. Lee JH, Reed DR, Price RA. Familial risk ratios for extreme obesity: implications for mapping human obesity genes. *Int. J. Obes. Relat. Metab. Disord.* 1997; 21:935–940.
10. Katzmarzyk PT, Pérusse L, Rao DC, Bouchard C. Familial risk of obesity and central adipose tissue distribution in the general Canadian population. *Am. J. Epidemiol.* 1999; 149:933–942.
11. Katzmarzyk PT, Pérusse L, Rao DC, Bouchard C. Familial risk of overweight and obesity in the Canadian population using the WHO/NIH criteria. *Obes. Res.* 2000; 8:194–197.
12. Pérusse L, Leblanc C, Bouchard C. Inter-generation transmission of physical fitness in the Canadian population. *Can. J. Sport. Sci.* 1988; 13:8–14.

13. Bouchard C, Pérusse L. Heredity and body fat. *Annu. Rev. Nutr.* 1988; 8:259–277.
14. Bouchard C, Després JP, Mauriege P, et al. The genes in the constellation of determinants of regional fat distribution. *Int. J. Obes.* 1991; 15:9–18.
15. Roberts SB, Greenberg AS. The new obesity genes. *Nutr. Rev.* 1996; 1:41–49.
16. Maes HH, Neale MC, Eaves LJ. Genetic and environmental factors in relative body weight and human adiposity. *Behav. Genet.* 1997; 27:325–351.
17. Bouchard C, Pérusse L, Rice T, Rao DC. The genetics of obesity. In: Bray G, Bouchard C, James W, eds. *Handbook of Obesity.* New York: Dekker, 1998:157–190.
18. Pérusse L, Chagnon YC, Rice T, Rao DC, Bouchard C. L'épidémiologie génétique et la génétique moléculaire de l'obésité: les enseignements de l'étude des familles de Québec. *Médecine Sciences* 1998; 14:914–924.
19. Pérusse L, Chagnon YC, Bouchard C. Etiology and genetics of massive obesity. In: Deitel M, Cowan GSM, eds. *Update: Surgery for the Morbidly Obese Patient.* Toronto: FD-Communications Inc., 2000:1–12.
20. Faith MS, Pietrobelli A, Nunez C, Heo M, Heymsfield SB, Allison DB. Evidence for independent genetic influences on fat mass and body mass index in a pediatric twin sample. *Pediatrics* 1999; 104:61–67.
21. Allison DB, Kaprio J, Korkeila M, Koskenvuo M, Neale MC, Hayakawa K. The heritability of body mass index among an international sample of monozygotic twins reared apart. *Int. J. Obes. Relat. Metab. Disord.* 1996; 20:501–506.
22. Sorensen TI, Holst C, Stunkard AJ. Adoption study of environmental modifications of the genetic influences on obesity. *Int. J. Obes. Relat. Metab. Disord.* 1998; 22:73–81.
23. Katzmarzyk PT, Perusse L, Bouchard C. Genetics of abdominal visceral fat levels. *Am. J. Human Biol.* 1999; 11:225–235.
24. Pérusse L, Després JP, Lemieux S, Rice T, Rao DC, Bouchard C. Familial aggregation of abdominal visceral fat level: results from the Quebec Family Study. *Metabolism* 1996; 45:378–382.
25. Rice T, Després JP, Daw EW, et al. Familial resemblance for abdominal visceral fat: the HERITAGE family study. *Int. J. Obes.* 1997; 21:1024–1031.
26. Pérusse L, Rankinen T, Zuberi A, et al. The human obesity gene map: the 2004 update. *Obes. Res.* 2005; 13:381–490.
27. Hill JO, Melanson EL. Overview of the determinants of overweight and obesity: current evidence and research issues. *Med. Sci. Sports Exerc.* 1999; 31:S515–521.
28. Lobstein T, Baur L, Uauy R. Obesity in children and young people: a crisis in public health. *Obes. Rev.* 2004; 5 Suppl 1:4–104.
29. Gortmaker SL, Must A, Sobol AM, Peterson K, Colditz GA, Dietz WH. Television viewing as a cause of increasing obesity among children in the United States, 1986–1990. *Arch. Pediatr. Adolesc. Med.* 1996; 150:356–362.
30. Marshall SJ, Biddle SJ, Gorely T, Cameron N, Murdey I. Relationships between media use, body fatness and physical activity in children and youth: a meta-analysis. *Int. J. Obes. Relat. Metab. Disord.* 2004; 28:1238–1246.
31. Hill JO, Melanson EL, Wyatt HT. Dietary fat intake and regulation of energy balance: implications for obesity. *J. Nutr.* 2000; 130:284S–288S.
32. Hunter DJ. Gene–environment interactions in human diseases. *Nat. Rev. Genet.* 2005; 6:287–298.
33. Pérusse L, Bouchard C. Role of genetic factors in childhood obesity and in susceptibility to dietary variations. *Ann. Med.* 1999; 31 Suppl 1:19–25.

34. Bouchard C, Perusse L, Leblanc C. Using MZ twins in experimental research to test for the presence of a genotype-environment interaction effect. *Acta. Genet. Med. Gemellol. (Roma).* 1990; 39:85–89.

35. Heitmann BL, Lissner L, Sorensen TI, Bengtsson C. Dietary fat intake and weight gain in women genetically predisposed for obesity. *Am. J. Clin. Nutr.* 1995; 61:1213–1217.

36. Heitmann BL, Kaprio J, Harris JR, Rissanen A, Korkeila M, Koskenvuo M. Are genetic determinants of weight gain modified by leisure-time physical activity? A prospective study of Finnish twins. *Am. J. Clin. Nutr.* 1997; 66:672–678.

37. Bouchard C, Tremblay A, Després JP, et al. The response to long-term overfeeding in identical twins. *N. Eng. J. Med.* 1990; 322:1477–1482.

38. Bouchard C, Tremblay A, Després J-P, et al. The response to exercise with constant energy intake in identical twins. *Obes. Res.* 1994; 2:400–410.

39. Ukkola O, Bouchard C. Role of candidate genes in the responses to long-term overfeeding: review of findings. *Obes. Rev.* 2004; 5:3–12.

40. Ukkola O, Kesaniemi YA, Tremblay A, Bouchard C. Two variants in the resistin gene and the response to long-term overfeeding. *Eur. J. Clin. Nutr.* 2004; 58:654–659.

41. Fumeron F, Durack-Bown I, Betoulle D, et al. Polymorphisms of uncoupling protein (UCP) and beta 3 adrenoreceptor genes in obese people submitted to a low calorie diet. *Int. J. Obes. Relat. Metab. Disord.* 1996; 20:1051–1054.

42. Hauner H, Meier M, Jockel KH, Frey UH, Siffert W. Prediction of successful weight reduction under sibutramine therapy through genotyping of the G-protein beta3 subunit gene (GNB3) C825T polymorphism. *Pharmacogenetics.* 2003; 13:453–459.

43. Nieters A, Becker N, Linseisen J. Polymorphisms in candidate obesity genes and their interaction with dietary intake of n-6 polyunsaturated fatty acids affect obesity risk in a sub-sample of the EPIC-Heidelberg cohort. *Eur. J. Nutr.* 2002; 41:210–221.

44. Robitaille J, Despres JP, Perusse L, Vohl MC. The PPAR-gamma P12A polymorphism modulates the relationship between dietary fat intake and components of the metabolic syndrome: results from the Quebec Family Study. *Clin. Genet.* 2003; 63:109–116.

45. Rankinen T, Perusse L, Rauramaa R, Rivera MA, Wolfarth B, Bouchard C. The human gene map for performance and health-related fitness phenotypes. *Med. Sci. Sports Exerc.* 2001; 33:855–867.

46. Rankinen T, Perusse L, Rauramaa R, Rivera MA, Wolfarth B, Bouchard C. The human gene map for performance and health-related fitness phenotypes: the 2001 update. *Med. Sci. Sports Exerc.* 2002; 34:1219–1233.

47. Pérusse L, Rankinen T, Rauramaa R, Rivera MA, Wolfarth B, Bouchard C. The human gene map for performance and health-related fitness phenotypes: the 2002 update. *Med. Sci. Sports Exerc.* 2003; 35:1248–1264.

48. Rankinen T, Perusse L, Rauramaa R, Rivera MA, Wolfarth B, Bouchard C. The human gene map for performance and health-related fitness phenotypes: the 2003 update. *Med. Sci. Sports Exerc.* 2004; 36:1451–1469.

49. Meirhaeghe A, Helbecque N, Cottel D, Amouyel P. Beta2-adrenoceptor gene polymorphism, body weight, and physical activity. *Lancet.* 1999; 353:896.

50. Corbalan MS. The 27Glu polymorphism of the beta2-adrenergic receptor gene interacts with physical activity influencing obesity risk among female subjects. *Clin. Genet.* 2002; 61:305–307.

51. Otabe S, Clement K, Dina C, et al. A genetic variation in the 5' flanking region of the UCP3 gene is associated with body mass index in humans in interaction with physical activity. *Diabetologia.* 2000; 43:245–249.

52. Garenc C, Perusse L, Chagnon YC, et al. Effects of beta2-adrenergic receptor gene variants on adiposity: the HERITAGE Family Study. *Obes. Res.* 2003; 11:612–618.

53. Phares DA, Halverstadt AA, Shuldiner AR, et al. Association between body fat response to exercise training and multilocus ADR genotypes. *Obes. Res.* 2004; 12:807–815.

54. Sakane N, Yoshida T, Umekawa T, Kogure A, Takakura Y, Kondo M. Effects of Trp64Arg mutation in the beta 3-adrenergic receptor gene on weight loss, body fat distribution, glycemic control, and insulin resistance in obese type 2 diabetic patients. *Diabetes Care.* 1997; 20:1887–1890.

55. Shiwaku K, Nogi A, Anuurad E, et al. Difficulty in losing weight by behavioral intervention for women with Trp64Arg polymorphism of the beta3-adrenergic receptor gene. *Int. J. Obes. Relat. Metab. Disord.* 2003; 27:1028–1036.

56. Lindi VI, Uusitupa MI, Lindstrom J, et al. Association of the Pro12Ala polymorphism in the PPAR-gamma2 gene with 3-year incidence of type 2 diabetes and body weight change in the Finnish Diabetes Prevention Study. *Diabetes.* 2002; 51:2581–2586.

57. Garenc C, Perusse L, Bergeron J, et al. Evidence of LPL gene-exercise interaction for body fat and LPL activity: the HERITAGE Family Study. *J. Appl. Physiol.* 2001; 91:1334–1340.

58. Tworoger SS, Chubak J, Aiello EJ, et al. The effect of CYP19 and COMT polymorphisms on exercise-induced fat loss in postmenopausal women. *Obes. Res.* 2004; 12:972–981.

59. Lanouette CM, Chagnon YC, Rice T, et al. Uncoupling protein 3 gene is associated with body composition changes with training in HERITAGE study. *J. Appl. Physiol.* 2002; 92:1111–1118.

60. Rankinen T, Rice T, Leon AS, et al. G protein beta 3 polymorphism and hemodynamic and body composition phenotypes in the HERITAGE Family Study. *Physiol. Genomics.* 2002; 8:151–157.

61. Peters WR, MacMurry JP, Walker J, Giese RJ, Jr., Comings DE. Phenylethanolamine N-methyltransferase G-148A genetic variant and weight loss in obese women. *Obes. Res.* 2003; 11:415–419.

62. Montgomery H, Clarkson P, Barnard M, et al. Angiotensin-converting-enzyme gene insertion/deletion polymorphism and response to physical training. *Lancet.* 1999; 353:541–545.

CHAPTER **8**

Saturated Fat Consumption in Ancestral Human Diets: Implications for Contemporary Intakes

Loren Cordain

CONTENTS

INTRODUCTION

Genetic Discordance

Nutritional requirements for all organisms are ultimately determined by the expression of specific genes within an organism's genome. These genes, in turn, are created and shaped by an ongoing interaction between the genome and its environment via evolution acting through natural selection over many generations. Genetic traits may be positively or negatively selected relative to their concordance or discordance with environmental selective pressures.[1] When the environment remains relatively constant over a long period of time, stabilizing selection tends to maintain genetic traits that represent the optimal average for a population.[2] On the other hand, when environmental conditions change over long periods of time, evolutionary discordance arises between a species' genome and its environment, and stabilizing selection is replaced by directional selection, moving the average population genome to a new set point.[1,2] Initially, when long term environmental changes occur in a population, individuals bearing the previous average status-quo genome experience evolutionary discordance.[2,3] In the affected genotype, this evolutionary discordance manifests itself phenotypically as disease, increased morbidity and mortality, and reduced reproductive success.[1–3]

Since the introduction of agriculture and animal husbandry 10,000 years ago, and more recently with the beginning of the Industrial Revolution 200 years ago, crucial changes have occurred in both diet and lifestyle conditions that are vastly different than the prevailing environmental conditions during which the human genome adapted. Numerous Neolithic and Industrial era food introductions have been identified that promote the development of chronic disease in contemporary western populations.[4–10] In most cases, a dose response exists between these novel foods and the emergence of disease. For instance, occasional seasonal exposure to honey (a refined sugar) results in negligible dental caries rates in hunter-gatherers,[11] whereas daily consumption of refined sucrose in Western diets almost universally causes a high incidence of caries and dental decay.[12] In many cases (such as with dental caries) the proximate physiological and biochemical causes for the diseases are well understood. Despite this knowledge, it is frequently less well appreciated that the ultimate basis for most diet-related diseases results from the evolutionary discordance between our ancient and conservative genome and recently introduced foods.[2,3,13] By examining pre-agricultural diets and their nutritional characteristics and comparing them to contemporary diets, insight can be gained into complex questions regarding diet and disease in existing populations.

Dietary Saturated Fats

A diet–disease question that has become contentious in recent years is saturated fats and the role they might play in the pathogenesis of coronary heart disease.[14–17] The traditional view has been that certain saturated fats (12:0, 14:0, and 16:0) downregulate the LDL receptor and thereby increase plasma concentrations of LDL cholesterol, which in turn increases the risk for coronary artery disease (CAD).[18,19] It is increasingly being recognized that this traditional model of atherosclerosis and CAD is overly simplistic, primarily because CAD is a multifactorial disease involving numerous dietary and genetic factors acting in concert with one another.[17] The dietary glycemic load, the n6/n3 fatty acid balance, chronic inflammation, trans fatty acids, homocysteine, alcohol intake, exercise, smoking, and numerous other dietary and lifestyle factors play key roles in the pathogenesis of CAD.[17] Nevertheless, the molecular[20] and clinical[21] basis for the elevation of plasma LDL by saturated fatty acids cannot be ignored, nor can the continuous and graded risk for CAD mortality with increasing LDL and total cholesterol concentrations,[22,23] despite suggestions otherwise.[14-16]

The relative contribution that dietary saturated fats may make to the overall development and progression of CAD under the backdrop of the typical Western diet and lifestyle is unclear, particularly given that individual genetic differences may modulate the cholesterol-raising effects of saturated fats.[17] However, this lack of precise evidence by no means exonerates dietary saturated fats. Rather, they represent a known risk factor for CAD that should be recognized and considered similar to other known dietary risk factors. In the current U.S. diet, an average of 11% of the daily energy is derived from saturated fat,[24] a figure slightly higher than the 10% or less recommended by the American Heart Association.[25] By examining the amounts of saturated fats in pre-agricultural hominin diets, an evolutionary baseline can be established regarding the normal range and limits of saturated fats that would have conditioned the human genome.

SATURATED FATS IN PRE-AGRICULTURAL DIETS

Figure 8.1 demonstrates that since the evolutionary emergence of hominins, 20 or more species may have existed.[26] Similar to historically studied hunter-gatherers,[27,28] there would have been no single, universal diet consumed by all extinct hominin species. Rather, diets would have varied by geographic locale, climate, and specific ecologic niche. However, a number of lines of evidence indicate that all hominin species and populations were omnivorous; consequently, dietary saturated fats would have always been a component in hominin diets.

Saturated Fat in Early Pliocene Hominin Diets

Our closest living primate relative, the chimpanzee (*Pan paniscus* and *Pan troglodytes*) is omnivorous and consumes a substantial amount of meat throughout the year obtained from hunting and scavenging.[29-31] Observational studies of wild chimpanzees demonstrate that during the dry season, meat intake is about 65 g per day for adults.[30] Accordingly, it is likely that the very earliest Pliocene hominins would have been capable of obtaining animal food through hunting and scavenging in a manner similar to chimpanzees. Additionally, fossils of early African hominins including *Australopithecus africanus,* and *Australopithecus robustu*s maintain carbon isotope signatures characteristic of omnivores.[32,33] Quantitative estimates of energy intake from animal food sources in these early hominins are unclear, other than that they were likely similar to, or greater than, estimated values (4 to 8.5% total energy) for chimpanzees.[30,34] Consequently, the amount of dietary saturated in the earliest hominin diets would have been substantially lower than later hominins whose diet became more dependent upon animal food energy sources.

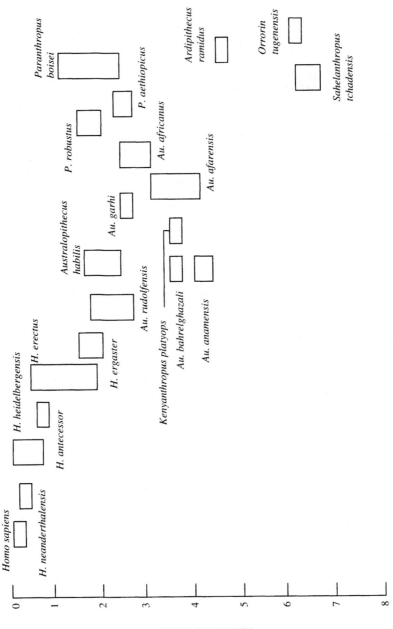

Figure 8.1   The hominin fossil record. Species are indicated with the dates of the earliest and latest fossil record. (Adapted from Wood, B., *Nature*, 418, 133, 2002.)

Saturated Fat in Pliocene/Pleistocene Hominin Diets

Approximately 2.6 million years ago (MYA), the hominin species that eventually led to *Homo* began to include more and more animal food in their diet. A number of lines of evidence support this viewpoint. First, Oldowan lithic technology appears in the fossil record 2.6 MYA,[35] and there is clear cut evidence to show that these tools were used to butcher and disarticulate animal carcasses.[36-37] Stone tool cut marks on the bones of prey animals and evidence for marrow extraction appear concurrently in the fossil record with the development of Oldowan lithic technology by at least 2.5 MYA.[37] It is not entirely clear which specific early hominin specie or species manufactured and used these earliest stone tools, however *Australopithecus garhi* might have been a likely candidate.[37,38]

The development of stone tools and the increased dietary reliance upon animal foods allowed early African hominins to colonize northern latitudes outside Africa where plant foods would have been seasonally restricted. Early *Homo* skeletal remains and Oldowan lithic technology appear at the Dmanisi site in the Republic of Georgia (40°N) by 1.75 MYA,[39] and more recently Oldowan tools dating to 1.66 MYA have been discovered at the Majuangou site in North China (40°N).[40] Both of these tool-producing hominins would likely have consumed considerably more animal food than pre-lithic hominins living in more temperate African climates because of reduced availability of plant foods during winter and early spring. Hence, the consumption of saturated fat would have, accordingly, been higher. Once again, quantitative estimates of the saturated fat content in early *Homo* species are speculative because of the uncertain plant–animal subsistence ratio. However, there is suggestive isotopic data indicating that the majority of the energy in more northerly living *Homo* species may have been obtained from animal foods.

Saturated Fat in Late Pleistocene Hominin Diets

Richards et al.[41] have examined stable isotopes (^{13}C and ^{15}N) in two Neanderthal specimens (~28,00 to 29,000 years BP) from Vindija Cave in northern Croatia and contrasted these isotopic signatures to those in fossils of herbivorous and carnivorous mammals from the same ecosystem. The analysis demonstrated that Neanderthals, similar to wolves and arctic foxes, behaved as top-level carnivores, obtaining all of their protein from animal sources. A comparable analysis was made of five Upper Paleolithic *Homo sapiens* specimens dated to the Upper Paleolithic (~11,700 to 12,380 years BP) from Gough's and Sun Hole Caves in Britain.[42] The data indicated these hunter-gatherers were consuming animal protein on a year-round basis at a higher trophic level than the arctic fox. Although precise quantitative estimates of saturated fat intake are not possible, the saturated fat intake in both Neanderthal and Upper Paleolithic *Homo sapiens* would have been substantial because of their great dependence upon animal food sources for daily energy.

Saturated Fat in Historically Studied Hunter-Gatherer Diets

Because reasonable estimates exist for the average plant-to-animal subsistence ratio for historically studied hunter-gatherers, it is possible to estimate the amount of saturated fat in their diet. Our analysis (Figure 8.2) of the *Ethnographic Atlas* data[43] showed that the dominant foods in the majority of historically studied hunter-gatherer diets were derived from animal food sources.[27] Most (73%) of the world's hunter-gatherers obtained >50% of their subsistence from hunted and fished animal foods, whereas only 14% of worldwide hunter-gatherers obtained >50% of their subsistence from gathered plant foods. For all 229 hunter-gatherer societies, the median subsistence dependence upon animal foods was 56 to 65%. In contrast, the median subsistence dependence upon gathered plant foods was 26 to 35%.[27]

The major limitation of ethnographic data is that the preponderance of it is subjective in nature, and the assigned scores for the five basic subsistence economies in the *Ethnographic Atlas* are not precise but, rather, are approximations.[44] Fortunately, more exact, quantitative dietary studies were carried out on a small percentage of the world's hunter-gatherer societies. Table 8.1 lists these studies and shows the plant to animal subsistence ratios by energy.[28] The average score for animal food subsistence is 65%, while that for plant food subsistence is 35%. When the two polar hunter-gatherer populations, who have no choice but to eat animal food because of the inaccessibility of plant food, are excluded from Table 8.1, the mean score for animal subsistence is 59% and that for plant food subsistence is 41%. These animal-to-plant subsistence values fall within the same respective class intervals (56 to 65% for animal food; 26 to 35% for plant food) as those we estimated from the ethnographic data when the confounding influence of latitude was eliminated.[27] Consequently, there is remarkably close agreement between the quantitative data in Table

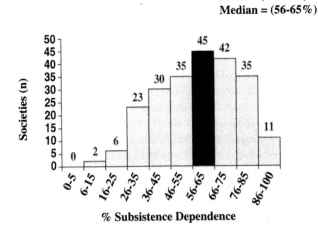

Figure 8.2 Frequency distribution of subsistence dependence upon total (fished + hunted) animal foods in world wide hunter-gatherer societies (n = 229). (Adapted from Cordain, L. et al., *Am. J. Clin. Nutr.*, 71, 682, 2000.)

Table 8.1   **Quantitatively Determined Proportions of Plant and Animal Food in Hunter-Gatherer Diets**

Population	Location	Latitude	% animal food	% plant food
Aborigines (Arhem Land)	Australia	12S	77	23
Ache	Paraguay	25S	78	22
Anbarra	Australia	12S	75	25
Efe	Africa	2N	44	56
Eskimo	Greenland	69N	96	4
Gwi	Africa	23S	26	74
Hadza	Africa	3S	48	52
Hiwi	Venezuela	6N	75	25
!Kung	Africa	20S	33	67
!Kung	Africa	20S	68	32
Nukak	Columbia	2N	41	59
Nunamiut	Alaska	68N	99	1
Onge	Andaman Islands	12N	79	21

Source: Cordain, L. et al., *Eur. J. Nutr.*, 56 (Suppl 1), S42, 2002.

8.1 and the ethnographic data that animal food comprised more than half of the energy in historically studied hunter-gatherer diets.

THE ESTIMATION OF DAILY DIETARY SATURATED FATS

Using the same model we developed for estimating the macronutrient content in hunter-gatherer diets,[27] it is possible to estimate the dietary saturated fat content, provided saturated fat values in the plant and animal food databases are known. Similar to our previous model, a range of plant-to-animal subsistence ratios are utilized to estimate the most likely range for dietary saturated fat.

Saturated Fat in Plant Foods

In the current model, fat contributed 24% of the total energy derived from all wild plant food (n = 768), whereas carbohydrate (62% energy) and protein (14%) comprised the balance of plant food energy. The mean fatty acid breakdown for 64 cultivated equivalent category plant foods was 22.4% saturated fatty acids, 28.6% monounsaturated fatty acids, and 49% polyunsaturated fatty acids.[45] Accordingly, in our model, 5.4% of plant food energy was derived from saturated fat.

Saturated Fat in Animal Foods

The estimation of saturated fat from animal sources is more complex because hunter-gatherers typically ate the entire edible carcass of most vertebrates,[46,47] thereby necessitating the calculation of the total edible carcass saturated fatty acid content. In mammals and most vertebrates, organ and tissue mass scales closely with body mass. Consequently, the mass of individual edible organs can be calculated from body mass using allometric equations.[48-51] The edible carcass mass can then be determined by subtracting the mass of the bones (minus marrow), hide, hooves, antlers, blood, urine, and gastrointestinal contents from the total live weight. Edible carcass saturated fatty acid mass can be computed by multiplying individual tissue and organ mass by their respective saturated fatty acid compositions (% mass) and then summing these values. Finally, the edible carcass saturated fatty acid content by energy can be calculated from values by mass using the cubic regression equations developed by Cordain et al.[27] Figure 8.3 shows the cubic relationship between edible body fat percent by mass and edible body saturated fat percent by energy in mammals. Application of this equation along with the saturated fat content of plant and fish foods, as previously described,[27] allows for the estimation of total dietary saturated fat when the relative plant-to-animal subsistence values are known (Table 8.2). In the current model, a range of likely plant-to-animal subsistence values in hunter-gatherer diets have been employed as previously outlined.[27] Note that in the current model, saturated fat content for fish was derived from the mean value (26.1% of total fat energy) from 20 species of fish.[45]

DISCUSSION

In Table 8.2 the mean dietary saturated fat as a percentage of total energy is 11.0 ± 3.9 (S.D.). However, it is likely that a number of the projected values are physiologically unrealistic because they encroach upon or exceed the physiologic protein ceiling.[27,52] If those values whose protein intake exceeds 35.1% of total energy are excluded from the analysis, the mean dietary saturated fat as a percentage of total energy is 13.2 ± 2.8. In the typical hunter-gatherer diet, the animal subsistence falls between 55 to 65% of total energy; consequently, in this group, the mean dietary

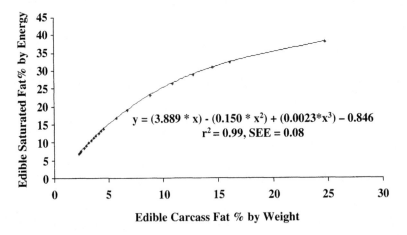

$$y = (3.889 * x) - (0.150 * x^2) + (0.0023*x^3) - 0.846$$
$$r^2 = 0.99, SEE = 0.08$$

Figure 8.3 Regression of whole edible carcass fat percentage by weight on edible carcass saturated fat % by energy.

Table 8.2   Dietary Macronutrient (% energy) and Saturated Fat (SAT) (% Energy) Estimates in Worldwide Hunter-Gatherer Societies (n = 229) with Varying Plant:Animal Subsistence Ratios and with Varying Animal (Hunted + Fished) Body Compositions

	% PRO%	% CHO	% FAT	% SAT FAT
(Plant:Animal) Subsistence Ratio				
35:65 – 20% animal fat	21	22	58	17.6
35:65 – 15% animal fat	28[a]	22	50	16.3
35:65 – 10% animal fat	35[a]	22	43	14.1
35:65 – 5% animal fat	47[a,b,c]	22	32	10.6
35:65 – 2.5% animal fat	56[a,b,c]	22	23	8.1
45:55 – 20% animal fat	20	28	52	15.8
45:55 – 15% animal fat	26	28	46	14.5
45:55 – 10% animal fat	32[a]	28	40	12.3
45:55 – 5% animal fat	42[a,b,c]	28	30	8.7
45:55 – 2.5% animal fat	49[a,b,c]	28	23	6.3
50:50 – 20% animal fat	20	31	49	15.1
50:50 – 15% animal fat	25	31	44	13.8
50:50 – 10% animal fat	31[a]	31	38	11.6
50:50 – 5% animal fat	39[a,b]	31	30	8.1
50:50 – 2.5% animal fat	46[a,b,c]	31	23	5.6
55:45 – 20% animal fat	19	34	47	14.5
55:45 – 15% animal fat	24	34	42	13.1
55:45 – 10% animal fat	29[a]	34	37	11.0
55:45 – 5% animal fat	37[a,b]	34	29	7.4
55:45 – 2.5% animal fat	43[a,b,c]	34	23	4.9
65:35 – 20% animal fat	19	40	41	13.1
65:35 – 15% animal fat	22	40	37	11.8
65:35 – 10% animal fat	26	40	34	9.6
65:35 – 5% animal fat	32[a]	40	28	6.1
65:35 – 2.5% animal fat	37[a,b]	40	23	3.6

[a] Exceeds low value (27.6% protein energy) for the range of maximal hepatic urea synthesis
[b] Exceeds mean value (35.1% protein energy) for the range of maximal hepatic urea synthesis
[c] Exceeds high value (40.9% protein energy) for the range of maximal hepatic urea synthesis

saturated fat as a percentage of total energy is higher still (15.1 ± 1.9). Even in plant dominated (>50% energy from plant foods) hunter-gatherer diets, the mean dietary saturated fat as a percentage of total energy is slightly higher (11.3 + 2.8) than recommended healthful values of <10%.[25]

The present data suggests that the normal dietary intake of saturated fatty acids that conditioned our species genome likely fell between 10 to 15% of total energy, and that values lower than 10% or higher than 15% would have been the exception rather than the rule. Consequently, population-wide recommendations to lower dietary saturated fat below 10% to reduce the risk of CAD[25] have little or no evolutionary foundation in pre-agricultural *Homo sapiens*. Because no randomized clinical trials of low saturated fat diets of sufficient duration have been carried out,[53] there is a lack of knowledge how low saturated fat intake can be without the risk of potentially deleterious health consequences.[17] Hence, extremely low, or conversely, high, lifelong consumption of dietary saturated fatty acids is likely to be discordant with the human genome.

REFERENCES

1. Gould, S.J., *The Structure Of Evolutionary Theory*, Harvard University Press, Cambridge, MA, 2002.
2. Boaz, N.T., *Evolving Health: The Origins Of Illness and How the Modern World Is Making Us Sick*, John Wiley & Sons, Inc., New York, 2002.
3. Nesse, R.M. and Williams G.C., *Why We Get Sick. The New Science of Darwinian Medicine*, Times Books, New York, 1994.
4. Eaton, S.B., Konner, M., and Shostak, M., Stone agers in the fast lane: chronic degenerative diseases in evolutionary perspective, *Am. J. Med.*, 84, 739, 1988.
5. Eaton, S.B., Eaton, S.B., III., and Konner, M.J., Paleolithic nutrition revisited: a twelve-year retrospective on its nature and implications, *Eur. J. Clin. Nutr.*, 51, 207, 1997.
6. O'Keefe, J.H., Jr. and Cordain, L., Cardiovascular disease resulting from a diet and lifestyle at odds with our Paleolithic genome: how to become a 21st-century hunter-gatherer, *Mayo Clin. Proc.*, 79, 101, 2004.
7. Cordain, L., The nutritional characteristics of a contemporary diet based upon Paleolithic food groups, *J. Am. Nutraceut. Assoc.*, 5, 15, 2002.
8. Sebastian, A., et al., Estimation of the net acid load of the diet of ancestral preagricultural Homo sapiens and their hominid ancestors, *Am. J. Clin. Nutr.*, 76, 1308, 2002.
9. Frassetto, L., et al., Diet, evolution and aging — the pathophysiologic effects of the post-agricultural inversion of the potassium-to-sodium and base-to-chloride ratios in the human diet, *Eur. J. Nutr.*, 40, 200, 2001.
10. Colagiuri, S. and Brand Miller, J., The "carnivore connection" — evolutionary aspects of insulin resistance, *Eur. J. Clin. Nutr.*, 56 Suppl 1, S30, 2002.
11. Turner, C.G., Dental anthropological indications of agriculture among the Jomon people of central Japan, *Am. J. Phys. Anthropol.*, 51, 619, 1979.
12. Zero, D.T., Sugars — the arch criminal? *Caries Res.*, 38, 277, 2004.
13. Eaton, S.B., et al., Evolutionary health promotion, *Prev. Med.*, 34, 109, 2002.
14. Ravnskov, U., et al., Studies of dietary fat and heart disease, *Science*, 295, 1464, 2002.
15. Ravnskov, U., The questionable role of saturated and polyunsaturated fatty acids in cardiovascular disease, *J. Clin. Epidemiol.*, 51, 443, 1998.
16. Taubes, G., The soft science of dietary fat, *Science*, 291, 2535, 2001.
17. German, J.B. and Dillard, C.J., Saturated fats: what dietary intake? *Am. J. Clin. Nutr.*, 80, 550, 2004.
18. Grundy, S.M., Dietary fat: at the heart of the matter, *Science*, 293, 801, 2001.
19. Katan, M.B., Zock, P.L., and Mensink, R.P., Dietary oils, serum lipoproteins, and coronary heart disease, *Am. J. Clin. Nutr.*, 61 (6 Suppl), 1368S, 1995.
20. Horton, J.D., Cuthbert, J.A., and Spady, D.K. Dietary fatty acids regulate hepatic low density lipoprotein (LDL) transport by altering LDL receptor protein and mRNA levels, *J. Clin. Invest.*, 92, 743, 1993.

21. Phinney, S.D., et al., The human metabolic response to chronic ketosis without caloric restriction: physical and biochemical adaptation, *Metabolism*, 32, 757, 1983.

22. Stamler, J., Wentworth, D., and Neaton, J.D., Is relationship between serum cholesterol and risk of premature death from coronary heart disease continuous and graded? Findings in 356,222 primary screenees of the Multiple Risk Factor Intervention Trial (MRFIT), *JAMA*, 256, 2823, 1986.

23. Stamler, J., et al., Relationship of baseline serum cholesterol levels in 3 large cohorts of younger men to long-term coronary, cardiovascular, and all-cause mortality and to longevity, *JAMA*, 284, 311, 2000.

24. Popkin, B.M., et al., Where's the fat? Trends in U.S. diets 1965–1996, *Prev. Med.*, 32, 245, 2001.

25. Krauss, R.M., et al., AHA Dietary Guidelines: revision 2000: a statement for healthcare professionals from the Nutrition Committee of the American Heart Association, *Circulation*, 102, 2284, 2000.

26. Wood, B., Palaeoanthropology: hominid revelations from Chad, *Nature*, 418, 133, 2002.

27. Cordain, L., et al., Plant to animal subsistence ratios and macronutrient energy estimations in world wide hunter-gatherer diets, *Am. J. Clin. Nutr.*, 71, 682, 2000.

28. Cordain, L., et al., The paradoxical nature of hunter-gatherer diets: Meat based, yet non-atherogenic, *Eur. J. Clin. Nutr.*, 56 (suppl 1), S42, 2002.

29. Schoeninger, M.J., Moore, J., and Sept, J.M., Subsistence strategies of two "savanna" chimpanzee populations: the stable isotope evidence, *Am. J. Primatol.*, 49, 297, 1999.

30. Stanford, C.B., The hunting ecology of wild chimpanzees: implications for the evolutionary ecology of Pliocene hominids, *Am. Anthropol.*, 98, 96, 1996.

31. Teleki, G., The omnivorous chimpanzee, *Sci. Am.*, 228, 33, 1973.

32. Lee-Thorp, J., Thackeray, J.F., and van der Merwe, N., The hunters and the hunted revisited, *J. Hum. Evol.*, 39, 565, 2000.

33. Sponheimer, M. and Lee-Thorp, J.A., 2003. Differential resource utilization by extant great apes and australopithecines: towards solving the C4 conundrum, *Comp. Biochem. Physiol. A Mol. Integr. Physiol.*, 136, 27, 2003.

34. Sussman, R.W., Foraging patterns of nonhuman primates and the nature of food preferences in man, *Fed. Proc.*, 37, 55, 1978.

35. Semaw, S., et al., 2.6-Million-year-old stone tools and associated bones from OGS-6 and OGS-7, Gona, Afar, Ethiopia, *J. Hum. Evol.*, 45, 169, 2003.

36. Bunn, H.T. and Kroll, E.M., Systematic butchery by Plio/Pleistocene hominids at Olduvai Gorge, Tanzania, *Curr. Anthropol.*, 17, 431, 1986.

37. de Heinzelin, J., et al., Environment and behavior of 2.5-million-year-old Bouri hominids, *Science*, 284, 625, 1999.

38. Asfaw, B., et al., Australopithecus garhi: a new species of early hominid from Ethiopia, *Science*, 284, 629, 1999.

39. Vekua, A., et al., A new skull of early Homo from Dmanisi, Georgia, *Science*, 297, 85, 2002.

40. Zhu, R.X., et al., New evidence on the earliest human presence at high northern latitudes in northeast Asia, *Nature*, 431, 559, 2004.

41. Richards, M.P., et al., Neanderthal diet at Vindija and Neanderthal predation: the evidence from stable isotopes, *Proc. Natl. Acad. Sci.*, 97, 7663, 2000.

42. Richards, M.P., et al., Focus: Gough's Cave and Sun Hole Cave human stable isotope values indicate a high animal protein diet in the British Upper Palaeolithic, *J. Archaeol. Sci.*, 27, 1, 2000.

43. Gray, J.P., A corrected ethnographic atlas, *World Cultures J.*, 10, 24, 1999.

44. Hayden, B., Subsistence and ecological adaptations of modern hunter/gatherers, in *Omnivorous Primates*, Harding, R.S.O. and Teleki, G., Eds., Columbia University Press, New York, 1981, 344.

45. Nutritionist V Nutrition Software (Version 2.3), N-Squared Computing, San Bruno, CA, 2000.

46. Thomas, E.M., *The Harmless People*, New York, Knopf, 1959.

47. McArthur, M., Food consumption and dietary levels of groups of aborigines living on naturally occurring foods, in *Records of the American-Australian Scientific Expedition to Arnhem Land*, Mountford C.P., Ed, Melbourne University Press, Melbourne, 1960, 90.

48. Stahl, W.R., Organ weights in primates and other mammals, *Science*, 150, 1039, 1965.

49. Calder, W.A., *Size, Function and Life History*, Harvard University Press, Cambridge, 1984.

50. Meadows, S.D. and Hakonson, T.E., Contributions of tissues to body mass in elk, *J. Wildl. Manage.*, 46, 838, 1982.

51. Hakonson, T.E. and Whicker, F.W., The contribution of various tissues and organs to total body mass in mule deer, *J. Mammal.*, 52, 628, 1971.

52. Rudman, D., et al., Maximal rates of excretion and synthesis of urea in normal and cirrhotic subjects, *J. Clin. Invest.*, 52, 2241, 1973.

53. Sacks, F.M. and Katan, M., Randomized clinical trials on the effects of dietary fat and carbohydrate on plasma lipoproteins and cardiovascular disease, *Am. J. Med.*, 113 Suppl 9B, 13S, 2002.

CHAPTER 9

Plant-Based Diets and Prevention of Cardiovascular Disease: Epidemiologic Evidence

Frank B. Hu

CONTENTS

INTRODUCTION

Plant-based foods (e.g., fruits and vegetables, nuts, natural vegetable oils, and whole grains) are important components of traditional diets in Mediterranean and Asian regions.[1] Such diets have been associated with low rates of cardiovascular disease (CVD) and mortality in these regions. Recent data linking plant-based diets to reduced risk of CVD have prompted numerous epidemiologic and clinical studies[2] on the relation between dietary patterns and CVD. In this report we summarize epidemiologic research on plant-based foods and dietary patterns as they relate to the risk of CVD.

FRUITS AND VEGETABLES

Data suggest that many nutrients in fruits and vegetables (e.g., dietary fiber, folate, potassium, flavonoids, and antioxidant vitamins) are associated with reduced risk of CVD. These findings are consistent with studies showing an association between decreased risk of CVD and total intake of fruits and vegetables[3–10] (Figure 9.1). Knekt et al.[3] report an inverse relation between intake of fruits and vegetables and risk of coronary heart disease (CHD). Bazzano et al.[10] found a similar association between intake of foods high in flavonoids (e.g., apples, berries, and onions) and death from CHD. Similarly, Gillman et al. found that increased fruit and vegetable intake offered protection against stroke.[4]

In the Physicians' Health Study,[9] Liu et al. noted a significant association between increased intake of vegetables rich in carotenoids (such as broccoli, carrots, spinach, lettuce, yellow squash, and tomatoes) and reduced risk of CHD. Data from Joshipura et al. suggest a dose-response relation between intake of fruits and vegetables and risk of stroke and CHD,[6,7] with cruciferous vegetables, citrus fruit and juices, and vitamin C-rich fruits and vegetables offering the most protection against stroke, and green leafy vegetables the most protection against CHD (Table 9.1). However, higher consumption of potatoes and french fries were associated with a lightly increased risk of both CHD and stroke, probably because these foods induce high glycemic and insulinemic responses.[12] Other data link dietary patterns to risk of CVD. A Mediterranean diet high in fruits and vegetables and α-linoleic acid is known to reduce recurrence of myocardial infarction MI and mortality compared with a regular low-fat diet.[13] Dietary patterns high in fruits and vegetables are known to improve blood pressure.[14,15]

Conversely, several clinical trials indicate that high-dose supplementation with beta carotene offers no protective benefits against CVD.[16–19] Researchers have yet to determine whether antioxidants consumed as supplements havethe same effect as

Table 9.1 Multivariate RRs[a] of CHD or Stroke Comparing the Highest vs. the Lowest Quintiles of Fruit and Vegetable Intakes in the Pooled Analyses of the Nurses' Health Study and the Health Professionals' Follow-Up Study

Food	RR and 95% CIs	
	CHD	Stroke
All fruits	0.80 (0.69–0.92)	0.69 (0.52–0.91)
All vegetables	0.82 (0.71–0.94)	0.90 (0.68–1.18)
Total citrus fruits	0.88 (0.77–1.00)	0.72 (0.47–1.11)
Citrus fruit juices	1.06 (0.85–1.32)	0.65 (0.51–0.84)
Cruciferous vegetables	0.86 (0.75–0.99)	0.71 (0.55–0.93)
Green leafy vegetables	0.72 (0.63–0.83)	0.76 (0.58–0.99)
Vitamin C-rich fruits and vegetables	0.91 (0.79–1.04)	0.68 (0.52–0.89)
Legumes	1.06 (0.91–1.24)	1.03 (0.77–1.39)
Potatoes (including French fries)	1.15 (0.78–1.70)	1.18 (0.90–1.54)

[a] RRs are based on the comparison between the highest and lowest quintiles of intake for each item, adjusted for standard cardiovascular risk factors.[2]

those consumed as part of a dietary pattern. Also, the optimal doses and combinations of multiple antioxidants for benefits are yet to be determined.

NUTS

Several epidemiologic studies have shown an inverse association between consumption of nuts and decreased risk of CHD.[20-25] Fraser et al. found that relative risk (RR) of nonfatal MI was 0.52 for 5 times per week; corresponding RR for fatal MI was 0.62.[20] Other studies report similar findings.[26] Jiang et al. recently found that higher consumption of nuts and peanut butter was independently associated with a significant decrease in risk of type 2 diabetes in women.[27]

Nuts contain mostly monounsaturated and polyunsaturated fats. Numerous metabolic studies have demonstrated that diets high in nuts (peanuts, walnuts, or almonds) significantly lower LDL cholesterol and the ratio of total cholesterol to HDL cholesterol.[28] Most nuts are rich in arginine, the precursor of endothelium-derived relaxing factor, nitric oxide (NO). NO is a potent vasodilator that can inhibit platelet adhesion and aggregation. Cooke et al. suggest that the anti-atherogenic effect of nuts might be related in part to the arginine-NO pathway.[29] Increased intake of alpha-linoleic acid (e.g., via walnuts) has also been associated with reduced risk of CHD,[30] as have other micronutrients found in nuts (i.e., magnesium, copper, folic acid, potassium, fiber, and vitamin E).

WHOLE GRAINS

Whole grain products (e.g., whole wheat breads, brown rice, oats, and barley) usually have lower glycemic index (GI) values[12] and are known to be rich in fiber, antioxidant vitamins, magnesium, and phytochemicals. Refined grain products lose substantial amounts of dietary fiber, vitamins, minerals, essential fatty acids, and phytochemicals during processing. Several studies suggest an inverse association between consumption of whole grain foods and risk of CVD.[20,31-33]

Jacobs et al. reported a relation between whole-grain intake and reduced risk of death from ischemic heart disease in postmenopausal women; between the first and last quintiles, RRs declined from 1.0 to 0.70, respectively (p = 0.02).[31] Liu et al. observed that women who ate nearly three servings of whole grains per week had a 25% lower risk of CHD compared with those who ate less than one serving per week.[32]

Data suggest that increased consumption of whole grains reduces risk of ischemic stroke independent of known CVD risk factors.[33] Evidence indicates that additional factors provide protective effects beyond those mediated through folate, fiber, and vitamin E. Cleveland et al. report that the average American eats less than one serving per day of whole grains.[34] Increasing consumption of whole grains represents an opportunity to reduce incidence of CVD.

DIETARY PATTERNS

Although numerous studies have examined the relation between intake of individual nutrients or foods and risk of CHD, less research has been done on the effects of dietary patterns.[35] Two major dietary patterns, prudent and Western, have been identified.[36,37] The former is characterized by higher intakes of fruits, vegetables, legumes, fish, poultry, and whole grains; the latter with higher intakes of red and processed meats, sweets and desserts, french fries, and refined grains (Figure 9.1).

A comparison of the highest and lowest quintiles in the Nurses' Health Study[36] showed a RR for CHD of 0.76 (p = 0.03) for the prudent diet, and 1.46 (p = 0.02) for the Western diet. These data are consistent with those from the Health Professionals' Follow-Up Study.[37] Relative risks from the lowest to the highest quintiles of the prudent diet ranged from 1.0 to 0.70, respectively (p = 0.0009). Relative risks for the Western diet ranged from 1.0 in the lowest quintile to 1.64 in the highest (p<0.0001).[36]

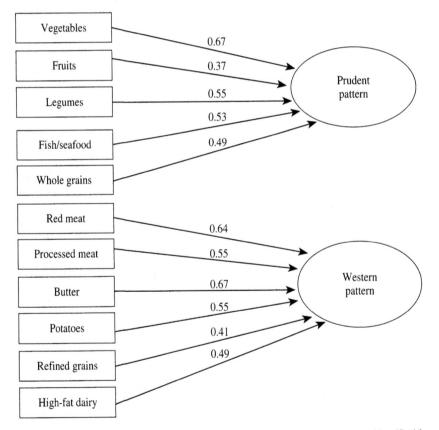

Figure 9.1   Factor loadings for selected foods in the two major dietary patterns identified from food frequency questionnaire in a subsample of the Health Professionals' Follow-Up Study. (From Hu, F.B., *Am. J. Clin. Nutr.* 78:544S–551S, 2003.)

Similarly, Fung et al. observed significant correlations between dietary patterns and plasma biomarkers of obesity and cardiovascular risk.[38] Jang et al. found that replacement of refined rice with whole grain and legume powder led to significant reductions in fasting glucose and insulin, homocysteine, and lipid peroxidation in patients with CVD.[39] These data are consistent with those from other clinical intervention studies.[14,40,41]

THE ROLE OF DIETARY FAT IN PLANT-BASED DIETS

Evidence suggests that the type of fat in plant-based diets is more important than the amount of fat in determining risk of CHD. Hu et al. found a significant positive association between intake of trans fatty acids and risk of CHD and an inverse association between polyunsaturated fat and CHD.[42] Data from several other prospective studies also showed a strong inverse relation between intake of polyunsaturated fats and CHD.[43,44,45] Metabolic studies have shown strong cholesterol-lowering effects of vegetable oils rich in linoleic acid.[46] Overall, diets high in polyunsaturated fats reduce total serum cholesterol and risk of CHD more effectively than do low-fat, high-carbohydrate diets.[30] These findings suggest that replacing saturated and trans fats with non-hydrogenated mono- and polyunsaturated fats might prevent CHD more effectively than reducing overall intake of fat.

Mediterranean countries in which olive oil (a major source of oleic acid) is the primary source of dietary fat have very low rates of CHD compared with most Western countries. These data suggest beneficial cardiovascular effects from monounsaturated fats, yet few data are available to support that hypothesis. Conversely, Posner et al. and Esrey et al. report a relation between higher intakes of monounsaturated fat and increased risk of CHD in younger subjects.[47,48] However, these studies did not adjust for other types of fats.

After adjusting for other kinds of fats, the ATBC Study found an inverse association between intake of monounsaturated fats and risk of CHD, as did the Nurses' Health Study.[42,49] Metabolic studies have shown that replacing carbohydrates with monounsaturated fat raises HDL without affecting LDL[50]; it also improves glucose tolerance and insulin sensitivity in patients with diabetes mellitus.[51] In addition, monounsaturated fat is resistant to oxidative modification.[52]

Major non-animal sources of monounsaturated fat include olive and canola oils, nuts, and avocados. Canola oil and nuts are also important sources of polyunsaturated fat. Natural liquid vegetable oils can be used in place of animal fat, stick margarine, and hydrogenated vegetable shortenings in cooking, frying, and baking. In general, vegetable oils contain higher amounts of vitamin E compared with animal fat, which has few antioxidants. Some vegetable oils (e.g., canola and soybean) are high in alpha-linolenic acid (ALA), an essential omega-3 fatty acid important to prevention of CVD. Data show that women who consumed oil and vinegar salad dressings (a major source of ALA) at least five to six times per week had approximately 50% lower risk of fatal CHD compared with those who rarely consumed this type of salad dressing.[53]

PLANT-BASED DIETS AND OBESITY

Although increased consumption of high-fat, energy-dense foods (e.g., nuts and vegetable oils) have raised concerns about weight gain, data show an inverse relation between intake of nuts and weight.[20,22] At the same time, there is no convincing evidence that reduced intake of dietary fat produces weight loss.[54] Today the role of carbohydrates in the development of obesity is receiving growing attention.[55] Experimental data suggest that diets with a high glycemic index (GI) contribute to obesity.[56,57] Denyer et al. report that high GI diets increase fat synthesis compared with isocaloric, low GI diets.[58] Roberts found a direct association between consumption of low GI foods and liquids and increased satiety.[57] High GI diets can contribute to weight gain and obesity by promoting excess energy consumption.[56] Evidence also suggests that diets lower in refined carbohydrates, with moderately high protein, improve blood lipids,[59,60] facilitate weight loss,[61] and lower risk of CHD.[62]

CONCLUSIONS

The past decade has seen an expansion of epidemiologic and clinical research on the roles of plant-based foods and dietary patterns in the prevention of CVD. This research has revolutionized our thinking about heart-healthy foods and the biological mechanisms that link dietary factors and CVD. Some foods that were thought unhealthy on the basis of fat content (e.g., nuts) have become important parts of diets designed to lower blood pressure[14] and serum cholesterol,[63] control weight,[64] enhance secondary prevention of CHD,[13] and add flavor, variety, and texture to meals. Other widely accepted foods (e.g., white bread and refined cereals) have been implicated in long-term adverse effects on insulin resistance and CHD.[55] Substantial evidence suggests that healthy plant-based diets — those with adequate omega-3 fatty acids that are rich in unsaturated fats, whole grains, fruits, and vegetables — can, and should, play an important part in the prevention of CVD and other chronic diseases.

ACKNOWLEDGMENTS

This work was supported by research grants HL34594, HL60712, and HL35464 from the National Institutes of Health. Dr.Hu is partly supported by an Established Investigator Award from the American Heart Association.

REFERENCES

1. Kushi LH, Lenart EB, and Willett WC. Health implications of Mediterranean diets in light of contemporary knowledge. 1. Plant foods and dairy products. *Am J Clin Nutr.* 1995;61 (suppl):1407S–1415S.

2. Hu FB. Plant-based foods and prevention of cardiovascular disease: an overview. *Am J Clin Nutr.* 2003;78:544S–551S.

3. Knekt P, Reunanen A, Jarvinen R, Seppanen R, Heliovaara M, and Aromaa A. Antioxidant vitamin intake and coronary mortality in a longitudinal population study. *Am J Epidemiol.* 1994;139:1180–1189.

4. Gillman MW, Cupples LA, Gagnon D, et al. Protective effect of fruits and vegetables on development of stroke in men. *J Am Med Assoc.* 1995;273:1113–1117.

5. Gaziano JM, Manson JE, Branch LG, Colditz GA, Willett WC, and Buring JE. A prospective study of consumption of carotenoids in fruits and vegetables and decreased cardiovascular mortality in the elderly. *Ann Epidemiol.* 1995;5:255–260.

6. Joshipura KJ, Ascherio A, Manson JE, et al. Fruit and vegetable intake in relation to risk of ischemic stroke. *JAMA.* 1999;282:1233–1239.

7. Joshipura KJ, Hu FB, Manson JE, et al. The effect of fruit and vegetable intake on risk for coronary heart disease. *Ann Intern Med.* 2001;134:1106–1114.

8. Liu S, Manson JE, Lee IM, et al. Fruit and vegetable intake and risk of cardiovascular disease: the Women's Health Study. *Am J Clin Nutr.* 2000;72:922–928.

9. Liu S, Lee IM, Ajani U, Cole SR, Buring JE, and Manson JE. Intake of vegetables rich in carotenoids and risk of coronary heart disease in men: The Physicians' Health Study. *Int J Epidemiol.* 2001;30:130–135.

10. Bazzano LA, He J, Odgden LG, et al. Fruit and vegetable intake and risk of cardiovascular disease in US adults: the first National Health and Nutrition Examination Survey Epidemiologic Follow-Up Study. *Am J Clin Nutr* 2002;76:93–99.

11. Knekt P, Jarvinen R, Reunanen A, and Maatela J. Flavonoid intake and coronary mortality in Finland: a cohort study. *Br Med J.* 1996;312:478–481.

12. Foster-Powell K and Miller JB. International tables of glycemic index. *Am J Clin Nutr.* 1995;62:871S–890S.

13. de Lorgeril M, Renaud S, Mamelle N, et al. Mediterranean alpha-linolenic acid-rich diet in secondary prevention of coronary heart disease. *Lancet.* 1994;343:1454–1459.

14. Appel L, Moore TJ, and Obrazanek E, for the DASH collaborative research group. A clinical trial of the effects of dietary patterns on blood pressure. *N Engl J Med.* 1997;336:1117–1124.

15. John JH, Ziebland S, Yudkin P, Roe LS, and Neil HAW, for the Oxford Fruit and Vegetable Study Group. Effects of fruit and vegetable consumption on plasma antioxidant concentrations and blood pressure: a randomized controlled trial. *Lancet.* 2002; http://image.thelancet.com/extras/01art9006web.pdf.

16. The Alpha-Tocopherol Beta-Carotene Cancer Prevention Study Group. The Alpha-Tocopherol, Beta-Carotene Lung Cancer Prevention Study: design, methods, participant characteristics, and compliance. *Ann Epidemiol.* 1994;4:1–10.

17. Greenberg ER and Sporn MB. Antioxidant vitamins, cancer, and cardiovascular disease. *N Engl J Med.* 1996;334:1189–1190.

18. Hennekens CH, Buring JE, Manson JE, et al. Lack of effect of long-term supplementation with beta carotene on the incidence of malignant neoplasms and cardiovascular disease. *N Engl J Med.* 1996;334:1145–1149.

19. Omenn GS, Goodman GE, Thornquist MD, et al. Effects of a combination of beta carotene and vitamin A on lung cancer and cardiovascular disease. *N Engl J Med.* 1996;334:1150–1155.

20. Fraser GE, Sabate J, Beeson WL, and Strahan TM. A possible protective effect of nut consumption on risk of coronary heart disease. The Adventist Health Study. *Arch Intern Med.* 1992;152:1416–1424.

21. Fraser GE and Shavlik DJ. Risk factors for all-cause and coronary heart disease mortality in the oldest-old. The Adventist Health Study. *Arch Intern Med.* 1997;157:2249–2258.

22. Hu FB, Stampfer MJ, Manson JE, et al. Frequent nut consumption and risk of coronary heart disease: prospective cohort study. *BMJ* 1998;317:1341–1345.

23. Brown L, Sacks F, Rosner B, and Willett WC. Nut consumption and risk of coronary heart disease in patients with myocardial infarction. *The FASEB J.* 1999;13(4):A4332.

24. Ellsworth JL, Kushi LH, and Folsom AR. Frequent nut intake and risk of death from coronary heart disease and all causes in postmenopausal women: the Iowa Women's Health Study. *Nutr Metab Cardiovasc Dis.* 2001;11:372–377.

25. Albert CM, Gaziano JM, Willett WC, and Manson JE. Nut consumption and decreased risk of sudden cardiac death in the Physicians' Health Study. *Arch Intern Med.* 2002;162:1382–1387.

26. Hu FB and Stampfer MJ. Nut consumption and risk of coronary heart disease: a review of epidemiologic evidence. *Current Atherosclerosis Reports.* 1999;I:204–209.

27. Jiang R, Manson JE, Stampfer MJ, Liu S, Willett WC, and Hu FB. Nut and peanut butter consumption and risk of type 2 diabetes in women. *JAMA.* 2002;288:2554–2560.

28. Kris-Etherton PM, Zhao G, Binkoski AE, Coval SM, and Etherton TD. The effects of nuts on coronary heart disease risk. *Nutr Rev.* 2001;59:103–111.

29. Cooke JP, Tsao P, Singer A, Wang B-Y, Kosek J, and Drexler H. Anti-atherogenic effect of nuts: Is the answer NO? *Arch Intern Med.* 1993;153:898–899 (letter).

30. Hu FB, Manson JE, and Willett WC. Types of dietary fat and risk of coronary heart disease: a critical review. *J Am Coll Nutr.* 2001;20:5–19.

31. Jacobs DR, Meyer KA, Kushi LH, and Folsom AR. Whole-grain intake may reduce the risk of ischemic heart disease death in postmenopausal women: the Iowa Women's Health Study. *Am J Clin Nutr.* 1998;68:248–257.

32. Liu S, Stampfer MJ, Hu FB, et al. Whole-grain consumption and risk of coronary heart disease: results from the Nurses' Health Study. *Am J Clin Nutr.* 1999;70:412–419.

33. Liu S, Manson JE, Stampfer MJ, et al. Whole grain consumption and risk of ischemic stroke in women: A prospective study. *JAMA.* 2000;284:1534–1540.

34. Cleveland LE, Moshfegh AJ, Albertson AM, Goldman JD. Dietary intake of whole grains (In Process Citation). *J Am Coll Nutr.* 2000;19:331S–338S.

35. Hu FB. Dietary pattern analysis: a new direction in nutritional epidemiology. *Curr Opin Lipidol.* 2002;13:3–9.

36. Fung TT, Willett WC, Stampfer MJ, Manson JE, and Hu FB. Dietary patterns and risk of coronary heart disease in women. *Arch Intern Med.* 2001;161:1857–1862.

37. Hu FB, Rimm EB, Stampfer MJ, Ascherio A, Spiegelman D, and Willett WC. Prospective study of major dietary patterns and risk of coronary heart disease in men. *Am J Clin Nutr.* 2000;72:912–921.

38. Fung TT, Rimm EB, Spiegelman D, et al. Association between dietary patterns and plasma biomarkers of obesity and cardiovascular disease risk. *Am J Clin Nutr.* 2001;73:61–67.

39. Jang Y, Lee JH, Kim OY, Park HY, and Lee SY. Consumption of whole grain and legume powder reduces insulin demand, lipid peroxidation, and plasma homocysteine concentrations in patients with coronary artery disease: randomized controlled clinical trial. *Arterioscler Thromb Vasc Biol.* 2001;21:2065–2071.

40. Jenkins DJ, Kendall CW, Faulkner D, et al. A dietary portfolio approach to cholesterol reduction: combined effects of plant sterols, vegetable proteins, and viscous fibers in hypercholesterolemia. *Metabolism.* 2002;51:1596–1604.

41. Appel LJ, Miller ER, 3rd, Jee SH, et al. Effect of dietary patterns on serum homocysteine: results of a randomized, controlled feeding study. *Circulation.* 2000;102:852–857.

42. Hu FB, Stampfer MJ, Manson JE, et al. Dietary fat intake and the risk of coronary heart disease in women. *N Engl J Med.* 1997;337:1491–1499.

43. Shekelle RB, Shryock AM, Paul O, et al. Diet, serum cholesterol, and death from coronary heart disease: The Western Electric Study. *N Engl J Med.* 1981;304:65–70.

44. Kushi LH, Lew RA, Stare FJ, et al. Diet and 20-year mortality from coronary heart disease: The Ireland-Boston Diet-Heart Study. *N Engl J Med.* 1985;312:811–818.

45. Dolecek TA. Epidemiological evidence of relationships between dietary polyunsaturated fatty acids and mortality in the multiple risk factor intervention trial. *Proc Soc Exp Biol Med.* 1992;200:177–182.

46. Kris-Etherton P, Yu S. Individual fatty acids on plasma lipids and lipoproteins: human studies. *Am J Clin Nutr.* 1997;65(suppl):1628S–1644S.

47. Posner BM, Cobb JL, Belanger AJ, Cupples A, D'Agostino RB, and Stokes III J. Dietary lipid predictors of coronary heart disease in men. *Arch Intern Med.* 1991;151:1181–1187.

48. Esrey KL, Joseph L, Grover SA. Relationship between dietary intake and coronary heart disease mortality: lipid research clinics prevalence follow-up study. *J. Clin Epidemiol.* 1996;2:211–216.

49. Pietinen P, Ascherio A, Korhonen P, et al. Intake of fatty acids and risk of coronary heart disease in a cohort of Finnish men: the ATBC Study. *Am J Epidemiol.* 1997;145:876–887.

50. Mensink RP and Katan MB. Effect of dietary fatty acids on serum lipids and lipoproteins: a meta-analysis of 27 trials. *Arterioscler and Thromb.* 1992;12:911–919.

51. Garg A. High-monounsaturated-fat diets for patients with diabetes mellitus: a meta-analysis. *Am J Clin Nutr.* 1998;67:563S–572S.

52. Parthasarathy S, Khoo JC, Miller E, Barnett J, Witztum JL, and Steinberg D. Low density lipoprotein rich in oleic acid is protected against oxidative modification: implications for dietary prevention of atherosclerosis. *Proc Natl Acad Sci USA.* 1990;87:3894–3898.

53. Hu FB, Stampfer MJ, Manson JE, et al. Dietary intake of alpha-linolenic acid and risk of iscehmic heart disease among women. *Am J Clin Nutr.* 1999;69:890–897.

54. Willett WC. Is dietary fat a major determinant of body fat? *Am J Clin Nutr.* 1998;67:556S–562S. (Published erratum appears in *Am J Clin Nutr* 1999 Aug;70 (2):304.)

55. Ludwig DS. The glycemic index: physiological mechanisms relating to obesity, diabetes, and cardiovascular disease. *Jama.* 2002;287:2414–23.

56. Ludwig DS. Dietary glycemic index and obesity. *J Nutr.* 2000;130:280S–283S.

57. Roberts SB. High-glycemic index foods, hunger, and obesity: is there a connection? *Nutr Rev.* 2000;58:163–169.

58. Denyer GS, Pawlak D, Higgins J, et al. Dietary carbohydrate and insulin resistance: lessons from human and animals. *Proceedings of the Nutrition Society of Australia.* 1998:158–167.

59. Wolfe BM. Potential role of raising dietary protein intake for reducing risk of atherosclerosis. *Can J Cardiol.* 1995;11 Supp G:127G–131G.

60. Jenkins DJ, Kendall CW, Vidgen E, et al. High-protein diets in hyperlipidemia: effect of wheat gluten on serum lipids, uric acid, and renal function. *Am J Clin Nutr.* 2001;74:57–63.
61. Skov AR, Toubro S, Ronn B, Holm L, and Astrup A. Randomized trial on protein vs carbohydrate in ad libitum fat reduced diet for the treatment of obesity. *Int J Obes Relat Metab Disord.* 1999;23:528–536.
62. Hu FB, Stampfer MJ, Manson JE, et al. Dietary protein and risk of coronary heart disease in women. *Am J Clin Nutr.* 1999;70:221–227.
63. Sabate J, Fraser GE, Burke K, Knutsen S, Bennett H, and Lindsted KD. Effects of walnuts on serum lipid levels and blood pressure in normal men. *N Engl J Med.* 1993;328:603–607.
64. McManus K, Antinoro L, and Sacks F. A randomized controlled trial of a moderate-fat, low-energy diet compared with a low fat, low-energy diet for weight loss in overweight adults. *Int J Obes Relat Metab Disord.* 2001;25:1503–1511.

CHAPTER **10**

Evolutionary Aspects of Diet, the Omega-6/Omega-3 Ratio, and Gene Expression

Artemis P. Simopoulos

CONTENTS

INTRODUCTION

The interaction of genetics and environment, nature and nurture, is the foundation for all health and disease. In the last two decades, using the techniques of molecular biology, it has been shown that genetic factors determine susceptibility to disease and environmental factors determine which genetically susceptible individuals will be affected.[1-6] Nutrition is an environmental factor of major importance. Whereas major changes have taken place in our diet over the past 10,000 years since the beginning of the Agricultural Revolution, our genes have not changed. The spontaneous mutation rate for nuclear DNA is estimated at 0.5% per million years. Therefore, over the past 10,000 years there has been time for very little change in our genes, perhaps 0.005%. In fact, our genes today are very similar to the genes of our ancestors during the Paleolithic period 40,000 years ago, at which time our genetic profile was established.[7] Genetically speaking, humans today live in a nutritional environment that differs from that for which our genetic constitution was selected. Studies on the evolutionary aspects of diet indicate that major changes have taken place in our diet, particularly in the type and amount of essential fatty acids and in the antioxidant content of foods[7-11] (Table 10.1; Figure 10.1). Using the tools of molecular biology and genetics, research is defining the mechanisms by which genes influence nutrient absorption, metabolism and excretion, taste perception, and degree of satiation — and the mechanisms by which nutrients influence gene expression.

Table 10.1 Characteristics of Hunter-Gatherer and Western Diet and Lifestyles

Characteristic	Hunter-Gatherer Diet and Lifestyle	Western Diet and Lifestyle
Physical Activity Level	high	low
Diet		
Energy density	low	high
Energy intake	moderate	high
Protein	high	low–moderate
Animal	high	low–moderate
Vegetable	very low	low–moderate
Carbohydrate	low–moderate (slowly absorbed)	moderate (rapidly absorbed)
Fiber	high	low
Fat	low	high
Animal	low	high
Vegetable	very low	moderate to high
Total long-chain ω6 + ω3	high (2.3 g/day)	low (0.2 g/day)
Ratio ω6/ω3	low (2.4)	high (12.0)
Vitamins, mg/d	**Paleolithic period**	**Current U.S. intake**
Riboflavin	6.49	1.34–2.08
Folate	0.357	0.149–0.205
Thiamin	3.91	1.08–1.75
Ascorbate	604	77–109
Carotene	5.56	2.05–2.57
(Retinol equivalent)	(927)	—
Vitamin A	17.2	7.02–8.48
(Retinol equivalent)	(2870)	(1170–429)
Vitamin E	32.8	7–10

Source: Modified from Simopoulos, A.P., in *Antioxidents in Nutrition and Health*, Papas, A., Ed., CRC Press, Boca Raton, FL, 1999.

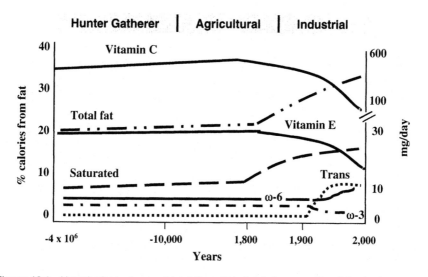

Figure 10.1   Hypothetical scheme of fat, fatty acid (ω6, ω3, trans, and total) intake (as percent of calories from fat), and intake of vitamins E and C (mg/d). Data were extrapolated from cross-sectional analyses of contemporary hunter-gatherer populations and from longitudinal observations and their putative changes during the preceding 100 years. (From Simopoulos, A.P., in *Antioxidents in Nutrition and Health*, Papas, A., Ed., CRC Press, Boca Raton, FL, 65,1999.)

EVOLUTIONARY ASPECTS OF DIET WITH EMPHASIS ON OMEGA-6 AND OMEGA-3 ESSENTIAL FATTY ACIDS

The foods that were commonly available to pre-agricultural humans (lean meat, fish, green leafy vegetables, fruits, nuts, berries, and honey) were the foods that shaped modern humans' genetic nutritional requirements. Cereal grains as a staple food are a relatively recent addition to the human diet and represent a dramatic departure from those foods to which we are genetically programmed and adapted.[12-17] Cereals did not become a part of our food supply until very recently — 10,000 years ago — with the advent of the Agricultural Revolution. Prior to the Agricultural Revolution humans ate an enormous variety of wild plants, whereas today about 17% of plant species provide 90% of the world's food supply, with the greatest percentage contributed by cereal grains.[12] Three cereals: wheat, maize, and rice together account for 75% of the world's grain production. Human beings have become entirely dependent upon cereal grains for the greater portion of their food supply. The nutritional implications of such a high grain consumption upon human health are enormous. Cereal grains are high in carbohydrates and omega-6 fatty acids, but low in omega-3 fatty acids and in antioxidants, particularly in comparison to green leafy vegetables. Recent studies show that low-fat, high-carbohydrate diets increase insulin resistance and hyperinsulinemia, conditions that increase the risk for coronary heart disease, hypertension, diabetes, and obesity.[18-21] And yet, for the 99.9% of mankind's presence on this planet, humans rarely or never consumed cereal grains. It is only since the last 10,000 years that humans have consumed cereals. Up to that time, humans were non-cereal-eating hunter-gatherers — since the emergence of Homo erectus 1.7 million years ago. There is no evolutionary precedent in our species for grass seed consumption.[7,12] Therefore, we have had little time (<500 generations) — since the beginning of the Agricultural Revolution 10,000 years ago — to adapt to a food type that now represents humanity's major source of both calories and protein. A number of anthropological, nutritional, and genetic studies indicate that humans' overall diet, including energy intake and energy expenditure, has changed over the past 10,000 years with major changes occurring during the past 150 years in the type and amount of fat and vitamins C and E intake[7,9,13,22,23,26,27] (Table 10.1 and Table 10.2; Figure 10.1).

Eaton and Konner[7] have estimated higher intakes for protein, calcium, potassium, and ascorbic acid and lower sodium intakes for the diet of the late Paleolithic period than the current U.S. and Western diets. Most of our food is calorically concentrated in comparison with wild game and the uncultivated fruits and vegetables of the Paleolithic diet. Paleolithic man consumed fewer calories and drank water, whereas today most drinks to quench thirst contain calories. Today industrialized societies are characterized by (1) an increase in energy intake and decrease in energy expenditure; (2) an increase in saturated fat, omega-6 fatty acids, and trans fatty acids, and a decrease in omega-3 fatty acid intake; (3) a decrease in complex carbohydrates and fiber; (4) an increase in cereal grains and a decrease in fruits and vegetables; and (5) a decrease in protein, antioxidants, and calcium intake[7,9,22,23,24,25,26] (Table 10.1–Table 10.3). The

Table 10.2 Estimated ω3 and ω6 Fatty Acid Intake in the Late Paleolithic Period (g/d)[a]

Plants	
LA	4.28
ALA	11.40
Animals	
LA	4.56
ALA	1.21
Total	
LA	8.84
ALA	12.60
Animal	
AA (ω6)	1.81
EPA (ω3)	0.39
DTA (ω6)	0.12
DPA (ω3)	0.42
DHA (ω3)	0.27
Ratios of ω6/ω3	
LA/ALA	0.70
AA+DTA/EPA+DPA+DHA	1.79
Total ω6/ω3	0.79[a]

Notes: LA, linoleic acid; ALA, α-linolenic acid; AA, arachidonic acid; EPA, eicosapentaenoic acid; DTA, docosatetraenoic acid; DPA, docosapentaenoic acid; DHA, docosahexaenoic acid.

[a] Assuming an energy intake of 35:65 of animal:plant sources.[23]

Source: Data from Eaton et al., World Rev. Nutr. Diet., 83, 12, 1998.

Table 10.3 Late Paleolithic and Currently Recommended Nutrient Composition for Americans

	Late Paleolithic	Current Recommendations
Total dietary energy, (%)		
Protein	33	12
Carbohydrate	46	58
Fat	21	30
Alcohol	~0	—
P/S ratio	1.41	1.00
Cholesterol (mg)	520	300
Fiber (g)	100–150	30–60
Sodium (mg)	690	1100–3300
Calcium (mg)	1500–2000	800–1600
Ascorbic acid (mg)	440	60

P/S = polyunsaturated to saturated fat

Source: Modified from Eaton et al., *World Rev. Nutr. Diet.,* 83, 12, 1998.

Table 10.4 Adverse Effects of Trans Fatty Acids

Increase
Low-density lipoprotein (LDL)
Platelet aggregation
Lipoprotein (a) [Lp(a)]
Body weight
Cholesterol transfer protein (CTP)
Abnormal morphology of sperm (in male rats)

Decrease or inhibit
Decrease or inhibit incorporation of other fatty acids into cell membranes
Decrease high-density lipoprotein (HDL)
Inhibit delta-6 desaturase (interfere with elongation and desaturation of essential fatty acids)
Decrease serum testosterone (in male rats)
Cross the placenta and decrease birth weight (in humans)

Source: Simopoulos, A.P., in *Obesity: New Directions in Assessment and Management,* VanItallie, T.B. and Simopoulos, A.P., Eds., Charles Press, Philadelphia, 1995, 241.

increase in trans fatty acids is detrimental to health as shown in Table 10.4.[28] In addition, trans fatty acids interfere with the desaturation and elongation of both omega-6 and omega-3 fatty acids, further decreasing the amount of arachidonic acid, eicosapentaenoic acid, and docosahexaenoic acid availability for human metabolism.[29]

BIOLOGICAL EFFECTS AND METABOLIC FUNCTIONS OF OMEGA-6 AND OMEGA-3 FATTY ACIDS

Food technology and agribusiness provided the economic stimulus that dominated the changes in the food supply.[30,31] From per capita quantities of foods available for consumption in the U.S. national food supply in 1985, the amount of eicosapentaenoic acid (EPA) is reported to be about 50 mg per capita/day and the amount of

Table 10.5 Omega-6:Omega-3 Ratios in Various Populations

Population	ω6/ω3	Reference
Paleolithic	0.79	[23]
Greece prior to 1960	1.00–2.00	[26]
Current Japan	4.00	[38]
Current India, rural	5–6.1	[39]
Current United Kingdom and northern Europe	15.00	[40]
Current United States	16.74	[23]
Current India, urban	38–50	[39]

DHA is 80 mg per capita/day. The two main sources are fish and poultry.[32] It has been estimated that the present Western diet is "deficient" in omega-3 fatty acids with a ratio of omega-6 to omega-3 of 15–20/1, instead of 1/1, as is the case with wild animals and, presumably, human beings.[7–11,23,33–35]

Thus, an absolute and relative change of omega-6 and omega-3 in the food supply of Western societies has occurred over the last 100 years. A balance existed between omega-6 and omega-3 for millions of years during the long evolutionary history of the genus *Homo*, and genetic changes occurred partly in response to these dietary influences. During evolution, omega-3 fatty acids were found in all foods consumed: meat, wild plants, eggs, fish, nuts, and berries. Recent studies by Cordain et al.[36] on wild animals confirm the original observations of Crawford and Sinclair et al.[33,37] However, rapid dietary changes over short periods of time as have occurred over the past 100–150 years is a totally new phenomenon in human evolution[23,26,38–40] (Table 10.5).

Mammalian cells cannot convert omega-6 to omega-3 fatty acids because they lack the converting enzyme, omega-3 desaturase. Linoleic acid (LA) and alpha-linolenic acid (ALA) and their long-chain derivatives are important components of animal and plant cell membranes. These two classes of essential fatty acids (EFA) are not interconvertible, are metabolically and functionally distinct, and often have important opposing physiological functions. The balance of EFA is important for good health and normal development. When humans ingest fish or fish oil, the EPA and docosahexaenoic acid (DHA) from the diet partially replace the omega-6 fatty acids, especially arachidonic acid (AA), in the membranes of probably all cells, but especially in the membranes of platelets, erythrocytes, neutrophils, monocytes, and liver cells.[8] Whereas cellular proteins are genetically determined, the polyunsaturated fatty acid (PUFA) composition of cell membranes is to a great extent dependent on the dietary intake. AA and EPA are the parent compounds for eicosanoid production (Table 10.6, Figure 10.2).

Because of the increased amounts of omega-6 fatty acids in the Western diet, the eicosanoid metabolic products from AA, specifically prostaglandins, thromboxanes, leukotrienes, hydroxy fatty acids, and lipoxins, are formed in larger quantities than those formed from omega-3 fatty acids, specifically EPA. The eicosanoids from AA are biologically active in very small quantities, and if they are formed in large amounts, they contribute to the formation of thrombus and atheromas; to allergic and inflammatory disorders, particularly in susceptible people; and to proliferation

Table 10.6 Effects of Ingestion of EPA and DHA from Fish or Fish Oil

- Decreased production of prostaglandin E_2 (PGE_2) metabolites
- A decrease in thromboxane A_2, a potent platelet aggregator and vasoconstrictor
- A decrease in leukotriene B_4 formation, an inducer of inflammation, and a powerful inducer of leukocyte chemotaxis and adherence
- An increase in thromboxane A_3, a weak platelet aggregator and weak vasoconstrictor
- An increase in prostacyclin PGI_3, leading to an overall increase in total prostacyclin by increasing PGI_3 without a decrease in PGI_2, both PGI_2 and PGI_3 are active vasodilators and inhibitors of platelet aggregation
- An increase in leukotriene B_5, a weak inducer of inflammation and a weak chemotactic agent

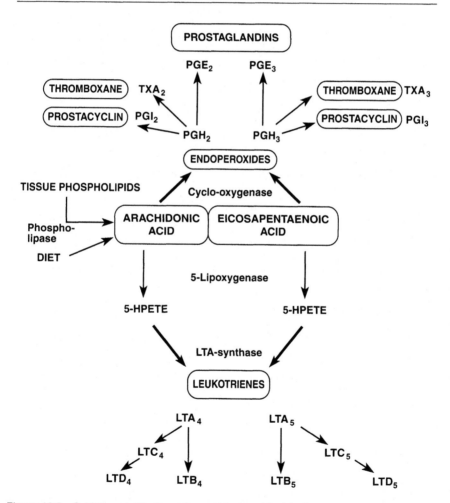

Figure 10.2 Oxidative metabolism of arachidonic acid and eicosapentaenoic acid by the cyclooxygenase and 5-lipoxygenase pathways. 5-HPETE denotes 5-hydroperoxyeicosatetranoic acid and 5-HPEPE denotes 5-hydroxyeicosapentaenoic acid.

of cells. Thus, a diet rich in omega-6 fatty acids shifts the physiological state to one that is prothrombotic and proaggregatory, with increases in blood viscosity, vasospasm, and vasoconstriction and decreases in bleeding time. Bleeding time is decreased in groups of patients with hypercholesterolemia,[41] hyperlipoproteinemia,[42] myocardial infarction, other forms of atherosclerotic disease, and diabetes (obesity and hypertriglyceridemia). Bleeding time is longer in women than in men and longer in young than in old people. There are ethnic differences in bleeding time that appear to be related to diet. Table 10.7 shows that the higher the ratio of omega-6/omega-3 fatty acids in platelet phospholipids, the higher the death rate from cardiovascular disease.[43]

The antithrombotic aspects and the effects of different doses of fish oil on the prolongation of bleeding time were investigated by Saynor et al.[44] A dose of 1.8 g/d EPA did not result in any prolongation in bleeding time, but at 4 g/d the bleeding time increased and the platelet count decreased without any adverse effects. In human studies there has never been a case of clinical bleeding, even in patients undergoing angioplasty while they were on fish oil supplements.[45]

There is substantial agreement that ingestion of fish or fish oils has the following effects: platelet aggregation to epinephrine and collagen is inhibited, thromboxane A_2 production is decreased, whole blood viscosity is reduced, and erythrocyte membrane fluidity is increased.[46–49] Fish oil ingestion increases the concentration of plasminogen activator and decreases the concentration of plasminogen activator inhibitor 1 (PAI-1).[50] *In vitro* studies have demonstrated that PAI-1 is synthesized and secreted in hepatic cells in response to insulin, and population studies indicate a strong correlation between insulinemia and PAI-1 levels. In patients with types IIb and IV hyperlipoproteinemia and in another double-blind clinical trial involving 64 men aged 35 to 40 years, ingestion of omega-3 fatty acids decreased the fibrinogen concentration.[51] Two other studies did not show a decrease in fibrinogen, but in one, a small dose of cod liver oil was used,[52] and in the other the study consisted of normal volunteers and was of short duration. A recent study noted that fish and fish oil increase fibrinolytic activity, indicating that 200 g/day of lean fish or 2 g of omega-3 EPA and DHA improve certain hematologic parameters implicated in the etiology of cardiovascular disease.[53]

Table 10.7 Ethnic Differences in Fatty Acid Concentrations in Thrombocyte Phospholipids and Percentage of All Deaths from Cardiovascular Disease

	Europe and United States	Japan	Greenland Eskimos
		%	
Arachidonic acid (20:4ω6)	26	21	8.3
Eicosapentaenoic acid (20:5ω3)	0.5	1.6	8.0
Ratio of ω6/ω3	50	12	1
Mortality from cardiovascular disease	45	12	7

Source: Data modified from Weber, P.C., Are we what we eat? Fatty acids in nutrition and in cell membranes: cell functions and disorders induced by dietary conditions, *Report No. 4*, Svanoy Foundation, Svanoybukt, Norway, 1989, 9.

Ingestion of omega-3 fatty acids not only increases the production of prosta-glandin I_3 (PGI_3), but also of PGI_2 in tissue fragments from the atrium, aorta, and saphenous vein obtained at surgery in patients who received fish oil two weeks prior to surgery.[54] Omega-3 fatty acids inhibit the production of platelet-derived growth factor (PDGF) in bovine endothelial cells.[55] PDGF is a chemoattractant for smooth muscle cells and a powerful mitogen. Thus, the reduction in its production by endothelial cells, monocytes/macrophages, and platelets could inhibit both the migration and proliferation of smooth muscle cells, monocytes/macrophages, and fibroblasts in the arterial wall. Insulin increases the growth of smooth muscle cells, leading to increased risk for the development of atherosclerosis. Omega-3 fatty acids increase endothelium-derived relaxing factor (EDRF).[56] EDRF (nitric oxide) facili-tates relaxation in large arteries and vessels. In the presence of EPA, endothelial cells in culture increase the release of relaxing factor, indicating a direct effect of omega-3 fatty acids on the cells.

Many experimental studies have provided evidence that incorporation of alter-native fatty acids into tissues may modify inflammatory and immune reactions and that omega-3 fatty acids, in particular, are potent therapeutic agents for inflammatory diseases. Supplementing the diet with omega-3 fatty acids (3.2 g EPA and 2.2 g DHA) in normal subjects increased the EPA content in neutrophils and monocytes more than sevenfold without changing the quantities of AA and DHA. The antiin-flammatory effects of fish oils are partly mediated by inhibiting the 5-lipoxygenase pathway in neutrophils and monocytes and inhibiting the leukotriene B_4 (LTB_4)-mediated function of LTB_5 (Figure 10.2).[57,58] Studies show that omega-3 fatty acids influence interleukin metabolism by decreasing IL-1β and IL-6.[59–62] Inflammation plays an important role in both the initiation of atherosclerosis and the development of atherothrombotic events.[63] An early step in the atherosclerotic process is the adhesion of monocytes to endothelial cells. Adhesion is mediated by leukocyte and vascular cell adhesion molecules (CAMs) such as selectins, integrins, vascular cell adhesion molecule 1 (VCAM-1), and intercellular adhesion molecule 1 (ICAM-1).[64] The expression of E-selectin, ICAM-1 and VCAM-1, which is relatively low in normal vascular cells, is upregulated in the presence of various stimuli, including cytokines and oxidants. This increased expression promotes the adhesion of mono-cytes to the vessel wall. The monocytes subsequently migrate across the endothelium into the vascular intima, where they accumulate to form the initial lesions of ath-erosclerosis. Atherosclerotic plaques have been shown to have increased CAM expression in animal models and human studies.[65–68] A balance between the omega-6 and omega-3 fatty acids is a more physiologic state in terms of gene expression,[69] eicosanoid metabolism, and cytokine production.

Further support for the need to balance the omega-6/omega-3 EFA comes from the studies of Ge et al.[70] and Kang et al.[71] The study by Ge et al. clearly shows the ability of both normal rat cardiomyocytes and human breast cancer cells in culture to form all the omega-3's from omega-6 fatty acids when fed the cDNA encoding omega-3 fatty acid desaturase obtained from the roundworm *C. elegans*. The omega-3 desaturase efficiently and quickly converted the omega-6 fatty acids that were fed to the cardiomyocytes in culture to the corresponding omega-3 fatty acids. Thus, omega-6 LA was converted to omega-3 ALA, and AA was converted to EPA, so that at equilibrium, the ratio of omega-6 to omega-3 PUFA was close to 1/1.[71] Further studies demonstrated that the cancer cells expressing the omega-3 desaturase underwent apoptotic death, whereas the control cancer cells with a high omega-6/omega-3 ratio continued to proliferate.[70] More recently, Kang et al. showed that transgenic mice expressing the *C. elegans fat-1* gene encoding an omega-3 fatty acid desaturase are capable of producing omega-3 from omega-6 fatty acids, leading to enrichment of omega-3 fatty acids with reduced levels of omega-6 fatty acids in almost all organs and tissues, including muscles and milk, with no need of dietary omega-3 fatty acid supply.[72] This discovery provides a unique tool and new opportunities for omega-3 research and raises the potential of production of *fat-1* transgenic livestock as a new and ideal source of omega-3 fatty acids to meet the human nutritional needs.

CLINICAL INTERVENTION STUDIES AND THE OMEGA-6/OMEGA-3 EFA BALANCE

The Lyon Heart Study was a dietary intervention study in which a modified diet of Crete (the experimental diet) was compared with the prudent diet, or Step I American Heart Association Diet (the control diet).[73–75] The experimental diet provided a ratio of LA to ALA of 4/1. This ratio was achieved by substituting olive oil and canola (oil) margarine for corn oil. Since olive oil is low in LA, whereas corn oil is high (8% and 61%, respectively) the ALA incorporation into cell membranes was increased. Cleland et al.[76] have shown that olive oil increases the incorporation of omega-3 fatty acids, whereas the LA from corn oil competes. The ratio of 4/1 of LA/ALA led to a 70% decrease in total mortality at the end of two years.[73]

The Gruppo Italiano per lo Studio della Sopravvivenza nell'Infarto miocardico (GISSI) Prevenzione Trial participants were on a traditional Italian diet plus 850 to 882 mg of omega-3 fatty acids at a ratio of 2/1 EPA to DHA.[77] The supplemented group had a decrease in sudden cardiac death by 45%. Although there are no dietary data on total intake for omega-6 and omega-3 fatty acids, the difference in sudden death is most likely due to the increase of EPA and DHA and a decrease of AA in cell membrane phospholipids. Prostaglandins derived from AA are proarrhythmic, whereas the corresponding prostaglandins from EPA are not.[78] In the Diet and Reinfarction Trial (DART), Burr et al. reported a decrease in sudden death in the group that received fish advice or took fish oil supplements relative to the group that did not.[79] Similar results have been obtained by Singh et al.[80,81] Studies carried out in India indicate that the higher ratio of 18:2ω6 to 18:3ω3 equalling 20/1 in their food supply led to increases in the prevalence of non-insulin dependent diabetes mellitus (NIDDM) in the population, whereas a diet with a ratio of 6/1 led to decreases.[82]

James and Cleland have reported beneficial effects in patients with rheumatoid arthritis[83] and Broughton has shown beneficial effects in patients with asthma by changing the background diet.[84] James and Cleland evaluated the potential use of omega-3 fatty acids within a dietary framework of an omega-6/omega-3 ratio of 3–4/1 by supplying 4 gm of EPA+DHA and using flaxseed oil rich in ALA. In their studies, the addition of 4 gm EPA and DHA in the diet produced a substantial inhibition of production of IL-1β and TNF when mononuclear cell levels of EPA were equal to or greater than 1.5% of total cell phospholipid fatty acids, which correlated with a plasma phospholipid EPA level equal to or greater than 3.2%. These studies suggest the potential for complementarity between drug therapy and dietary choices that increased intake of omega-3 fatty acids and decreased intake of omega-6 fatty acids may lead to drug-sparing effects. Therefore, future studies need to address the fat composition of the background diet, and the issue of concurrent drug use. A diet rich in omega-3 fatty acids and poor in omega-6 fatty acids provides the appropriate background biochemical environment in which drugs function.

Asthma is a mediator-driven inflammatory process in the lungs and the most common chronic condition in childhood. The leukotrienes and prostaglandins are implicated in the inflammatory cascade that occurs in asthmatic airways. There is evidence of airway inflammation even in newly diagnosed asthma patients within 2 to 12 months after their first symptoms.[85] Among the cells involved in asthma are mast cells, macrophages, eosinophils, and lymphocytes. The inflammatory mediators include cytokines and growth factors (peptide mediators) as well as the eicosanoids, which are the products of AA metabolism, which are important mediators in the underlying inflammatory mechanisms of asthma (Figure 10.2, Table 10.8). Leuko-trienes and prostaglandins appear to have the greatest relevance to the pathogenesis of asthma. The leukotrienes are potent inducers of bronchospasm, airway edema, mucus secretion, and inflammatory cell migration, all of which are important to the asthmatic symptomatology. Broughton et al.[84] studied the effect of omega-3 fatty acids at a ratio of omega-6/omega-3 of 10/1 to 5/1 in an asthmatic population in ameliorating methacholine-induced respiratory distress. With low omega-3 inges-tion, methacholine-induced respiratory distress increased. With high omega-3 fatty acid ingestion, alterations in urinary 5-series leukotriene excretion predicted treat-ment efficacy and a dose change in >40% of the test subjects (responders), whereas the non-responders had a further loss in respiratory capacity. A urinary ratio of 4-series to 5-series of <1 induced by omega-3 fatty acid ingestion may predict respi-ratory benefit.

Bartram et al.[86,87] carried out two human studies in which fish oil supplementation was given in order to suppress rectal epithelial cell proliferation and PGE_2 biosyn-thesis. This was achieved when the dietary omega-6/omega-3 ratio was 2.5/1 but not with the same absolute level of fish oil intake and an omega-6/omega-3 ratio of 4/1. More recently, Maillard et al. reported their results on a case control study.[88] They determined omega-3 and omega-6 fatty acids in breast adipose tissue and relative risk of breast cancer. They concluded, "our data based on fatty acid levels in breast adipose tissue (which reflect dietary intake) suggest a protective effect of omega-3 fatty acids on breast cancer risk and support the hypothesis that the balance between omega-3 and omega-6 fatty acids plays a role in breast cancer."

Psychologic stress in humans induces the production of proinflammatory cyto-kines such as interferon gamma (IFNγ), tumor necrosis factor α (TNFα), IL-6, and IL-10. An imbalance of omega-6 and omega-3 PUFA in the peripheral blood causes an overproduction of proinflammatory cytokines. There is evidence that changes in fatty acid composition are involved in the pathophysiology of major depression. Changes in serotonin (5-HT) receptor number and function caused by changes in

Table 10.8 Effects of Omega-3 Fatty Acids on Factors Involved in the Pathophysiology of Inflammation

Factor	Function	Effect of ω3 fatty acid
Arachidonic acid	Eicosanoid precursor; aggregates platelets; stimulates white blood cells	↓
Thromboxane	Platelet aggregation; vasoconstriction; increase of intracellular Ca^{++}	↓
Prostacyclin ($PGI_{2/3}$)	Prevent platelet aggregation; vasodilation; increase cAMP	↑
Leukotriene (LTB_4)	Neutrophil chemoattractant; increase of intracellular Ca^{++}	↓
Fibrinogen	A member of the acute phase response; and a blood clotting factor	↓
Tissue plasminogen activator	Increase endogenous fibrinolysis	↑
Platelet activating factor (PAF)	Activates platelets and white blood cells	↓
Platelet-derived growth factor (PDGF)	Chemoattractant and mitogen for smooth muscles and macrophages	↓
Oxygen free radicals	Cellular damage; enhance LDL uptake via scavenger pathway; stimulate arachidonic acid metabolism	↓
Lipid hydroperoxides	Stimulate eicosanoid formation	↓
Interleukin 1 and tumor necrosis factor	Stimulate neutrophil O_2 free radical formation; stimulate lymphocyte proliferation; stimulate PAF; express intercellular adhesion molecule-1 on endothelial cells; inhibit plasminogen activator, thus, procoagulants	↓
Interleukin-6	Stimulates the synthesis of all acute phase proteins involved in the inflammatory response: C-reactive protein; serum amyloid A; fibrinogen; α_1-chymotrypsin; and haptoglobin	↓

Source: Adapted and modified from Weber, P.C., and Leaf, A., in *Health Effects of w3 Polyunsaturated Fatty Acids in Seafoods*, Simopoulos, A.P. et al., Eds., vol. 66 *World Rev. Nutr. Diet*, Karger, Basel, 1991, 218.

PUFA provide the theoretical rationale connecting fatty acids with the current receptor and neurotransmitter theories of depression.[89–91] The increased C20:4ω6/C20:5ω3 ratio and the imbalance in the omega-6/omega-3 PUFA ratio in major depression may be related to the increased production of proinflammatory cytokines and eicosanoids in that illness.[89] There are a number of studies evaluating the therapeutic effect of EPA and DHA in major depression. Stoll and colleagues have shown that EPA and DHA prolong remission, that is, reduce the risk of relapse in patients with bipolar disorder.[92,93]

The above clinical studies in patients with cardiovascular disease, arthritis, asthma, cancer, and mental illness clearly indicate the need to balance the omega-6/omega-3 fatty acid intake for prevention and during treatment. The scientific evidence is strong for decreasing the omega-6 and increasing the omega-3 intake to improve health throughout the lifecycle.[94] The scientific basis for the development of a public policy to develop dietary recommendations for essential fatty acids, including a balanced omega-6/omega-3 ratio is robust.[95]

OMEGA-3 FATTY ACIDS AND GENE EXPRESSION

Previous studies have shown that fatty acids released from membrane phospholipids by cellular phospholipases, or made available to the cell from the diet or other aspects of the extracellular environment, are important cell-signalling molecules.[96] They can act as second messengers or substitute for the classical second messengers of the inositide phospholipid and the cyclic AMP signal transduction pathways.[96] They can also act as modulator molecules mediating responses of the cell to extracellular signals.[96] Recently it has been shown that fatty acids rapidly and directly alter the transcription of specific genes.[97]

Table 10.9 and Table 10.10 summarize the effects of various PUFA on gene expression. In the case of enzymes involved in carbohydrate and lipid metabolism, both omega-3 and omega-6 fatty acids appear to suppress the genes that encode for several enzymes (Table 10.9), whereas saturated, trans-, and monounsaturated fatty acids fail to suppress. DHA appears more potent in its effect than other PUFA. Omega-6 and omega-3 fatty acids and monounsaturated fatty acids induce acyl-CoA oxidase, the enzyme involved in beta-oxidation, but here again, DHA appears to be more potent.

In studies of inflammatory cytokines, such as IL-1β, both EPA and DHA suppress IL-1β mRNA whereas AA does not, and the same effect appears in studies on growth-related early response gene expression and growth factor (Table 10.10). In the case of VCAM, AA has a modest suppressing effect relative to DHA. The latter situation may explain the protective effect of fish oil toward colonic carcinogenesis, since EPA and DHA did not stimulate protein kinase C.[111] PUFA regulation of gene expression extends beyond the liver and includes genes such as adipocyte glucose transporter-4, lymphocyte stearoyl-CoA desaturase 2 in the brain, peripheral monocytes (IL-1β, and VCAM-1), and platelets (PDGF)[110] (Table 10.9 and Table 10.10). Whereas some of the transcriptional effects of PUFA appear to be mediated by eicosanoids, the PUFA suppression of lipogenic and glycolytic genes is independent of eicosanoid synthesis and appears to involve a nuclear mechanism directly modified by PUFA. Because of their coordinate or opposing effects, both classes of PUFA are needed in the proper amounts for normal growth and development. Although, so far the studies in infants have concentrated on the effects of PUFA on retinal and brain phospholipid composition and intelligence quotient (IQ),[112,113] motor development is very much dependent on intermediary metabolism and on overall normal metabolism, both of which are influenced by fatty acid biosynthesis and carbohydrate metabolism.

The amounts of PUFA found in breast milk in mothers fed diets consistent with our evolution should serve as a guide to determine omega-6 and omega-3 fatty acid requirements during pregnancy, lactation, and infant feeding. Of interest is the fact that saturated, monounsaturated, and trans fatty acids do not exert any suppressive action on lipogenic or glycolytic gene expression, which is consistent with their high

Table 10.9 Effects of Polyunsaturated Fatty Acids on Several Genes Encoding Enzyme Proteins Involved in Lipogenesis, Glycolysis, and Glucose Transport

Function and Gene	Reference	Linoleic acid	α-Linolenic acid	Arachidonic acid	Eicosapentaenoic acid	Docosahexaenoic acid
Hepatic cells						
Lipogenesis						
FAS	98–101	→	→	→	→	→
S14	98–101	→	→	→	→	→
SCD1	102	→	→	→	→	→
SCD2	103	→	→	→	→	→
ACC	101	→	→	→	→	→
ME	101	→	→	→	→	→
Glycolysis						
G6PD	104	→	→	→	→	→
GK	104	→	→	→	→	→
PK	105	—				
Mature adiposites						
Glucose transport						
GLUT4	106	—	—	→ ←	→ ←	—
GLUT1	106	—	—			—

Table 10.10 Effects of Polyunsaturated Fatty Acids on Several Genes Encoding Enzyme Proteins Involved in Cell Growth, Early Gene Expression, Adhesion Molecules, Inflammation, β-Oxidation, and Growth Factors[a]

Function and Gene	Reference	Linoleic acid	α-Linolenic acid	Arachidonic acid	Eicosapentaenoic acid	Docosahexaenoic acid
Cell growth and early gene expression						
c-fos	107	—	—	↑	→	→
Egr-1	107	—	—	↑	→	→
Adhesion molecules						
VCAM-1 mRNA[b]	108	—	—	→	c	→
Inflammation						
IL-1β	109	—	—	↑	→	→
β-oxidation						
Acyl-CoA oxidase[d]	101	↑	↑	↑	↑↑	↑
Growth factors						
PDGF	110	—	—	↑	→	→

a VCAM, vascular cell adhesion molecule; IL, interleukin; PDGF, platelet-derived growth factor. ↓ suppresses or decreases, ↑ induces or increases.
b Monounsaturated fatty acids (MONOs) also suppress VCAM1 mRNA, but to a lesser degree than does DHA. AA also suppresses to a lesser extent than DHA.
c Eicosapentaenoic acid has no effect by itself but enhances the effect of docosahexaenoic acid (DHA).
d MONOs also induce acyl-CoA oxidase mRNA.

content in human milk serving primarily as sources of energy. Because nutrients influence gene expression, and many chronic diseases begin in utero or in infancy, proper dietary intake of PUFA, even prior to pregnancy may be essential, as shown for folate deficiency in the development of neural tube defects.

DIET–GENE INTERACTIONS: GENETIC VARIATION AND OMEGA-6 AND OMEGA-3 FATTY ACID INTAKE IN THE RISK FOR CARDIOVASCULAR DISEASE

As discussed above, leukotrienes are inflammatory mediators generated from AA by the enzyme 5-lipoxygenase (Figure 10.2). Since atherosclerosis involves arterial inflammation, Dwyer et al. hypothesized that a polymorphism in the 5-lipoxygenase gene promoter could relate to atherosclerosis in humans, and that this effect could interact with the dietary intake of competing 5-lipoxygenase substrates.[114] The study consisted of 470 healthy middle-aged women and men from the Los Angeles Atherosclerosis Study, randomly sampled. The investigators determined 5-lipoxygenase genotypes, carotid-artery intima thickness, markers of inflammation, C-reactive protein (CRP), interleukin-6 (IL-6), dietary AA, EPA, DHA, LA, and ALA, with the use of six 24-hour recalls of food intake. The results showed that 5-lipoxygenase variant genotypes were found in 6% of the cohort. Mean intima-media thickness adjusted for age, sex, height, and racial or ethnic group was increased by 80 ± 19 μm from among the carriers of two variant alleles as compared with the carrier of the common (wild-type) allele. In multivariate analysis, the increase in intima-media thickness among carriers of two variant alleles (62 μm, p = 0.001) was similar in this cohort to that associated with diabetes (64 μm, p = 0.01), the strongest common cardiovascular risk factor. Increased dietary AA significantly enhanced the apparent atherogenic effect of genotype, whereas increased dietary intake of omega-3 fatty acids EPA and DHA blunted this effect. Furthermore, the plasma level of CRP of two variant alleles was increased by a factor of 2, as compared with that among carriers of the common allele. Thus, genetic variation of 5-lipoxygenase identifies a subpopulation with increased risk for atherosclerosis. The diet–gene interaction further suggests that dietary omega-6 fatty acids promote, whereas marine omega-3 fatty acids EPA and DHA inhibit, leukotriene-mediated inflammation that leads to atherosclerosis in this subpopulation.

The prevalence of variant genotypes did differ across racial and ethnic groups with higher prevalence among Asians or Pacific Islanders (19.4%), blacks (24%), and other racial or ethnic groups (18.2%) than among Hispanic subjects (3.6%) and non-Hispanic whites (3.1%). Increased intima-mediated thickness was significantly associated with intake of both AA and LA among carriers of the two variant alleles, but not among carriers of the common alleles. In contrast, the intake of marine omega-3 fatty acids was significantly and inversely associated with intima-media thickness only among carriers of the two variant alleles. Diet–gene interactions were specific to these fatty acids and were not observed for dietary intake of monounsaturated, saturated fat, or other measured fatty acids. The study constitutes evidence that genetic variation in an inflammatory pathway — in this case the leukotriene pathway — can trigger atherogenesis in humans. These findings could lead to new dietary and targeted molecular approaches to the prevention and treatment of cardiovascular disease according to genotype, particularly in the populations of non-European descent.

CONCLUSIONS

- Evidence from studies on the evolutionary aspects of diet, modern day hunter-gatherers, and traditional diets indicate that human beings evolved on a diet in which the ratio of omega-6/omega-3 EFA was about 1, whereas in the Western diets the ratio is 15/1 to 16.7/1. Agribusiness and modern agriculture have led to decreases in omega-3 fatty acids and increases in omega-6 fatty acids. Such practices have led to excessive amounts of omega-6 fatty acids, upsetting the balance that was characteristic during evolution when our genes were programmed to respond to diet and other aspects of the environment.
- LA and ALA are not interconvertible and compete for the rate-limiting $\Delta6$-desaturase in the synthesis of long-chain PUFA.
- AA (omega-6) and EPA (omega-3) are the parent compounds for the production of eicosanoids. Eicosanoids from AA have opposing properties from those of EPA. An increase in the dietary intake of omega-6 EFA changes the physiological state to a prothrombotic, proconstrictive, and proinflammatory state.
- Many of the chronic conditions — cardiovascular disease, diabetes, cancer, obesity, autoimmune diseases, rheumatoid arthritis, asthma, and depression — are associated with increased production of thromboxane A_2 (TXA_2), leukotriene B_4 (LTB_4), IL-1β, IL-6, tumor necrosis factor (TNF), and C-reactive protein. All these factors increase by increases in omega-6 fatty acid intake and decrease by increases in omega-3 fatty acid intake, either ALA or EPA and DHA. EPA and DHA are more potent, and most studies have been carried out using EPA and DHA.
- In the secondary prevention of cardiovascular disease, a ratio of 4/1 was associated with a 70% decrease in total mortality. A ratio of 2.5/1 reduced rectal cell proliferation in patients with colorectal cancer, whereas a ratio of 4/1 with the same amount of omega-3 PUFA had no effect. The lower omega-6/omega-3 ratio in women with breast cancer was associated with decreased risk. A ratio of 2–3/1 suppressed inflammation in patients with rheumatoid arthritis, and a ratio of 5/1 had a beneficial effect on patients with asthma, whereas a ratio of 10/1 had adverse consequences. These studies indicate that the optimal ratio may vary with the disease under consideration. This is consistent with the fact that chronic diseases are multigenic and multifactorial. Therefore, it is quite possible that the therapeutic dose of omega-3 fatty acids will depend on the degree of severity of disease resulting from the genetic predisposition.
- Studies show that the background diet, when balanced in omega-6/omega-3, decreases the drug dose. It is therefore essential to decrease the omega-6 intake while increasing the omega-3 in the prevention and management of chronic disease. Furthermore, the balance of omega-6 and omega-3 fatty acids is very important for homeostasis and normal development. The ratio of omega-6 to omega-3 EFA is an important determinant of health, because both omega-6 and omega-3 fatty acids influence gene expression. Recent studies on diet–gene interaction further suggest that dietary omega-6 fatty acids promote, whereas marine omega-

3 fatty acids EPA and DHA inhibit leukotriene-mediated inflammation that leads to atherosclerosis.

REFERENCES

1. Simopoulos, A.P. and Childs, B., Eds., *Genetic Variation and Nutrition,* vol. 63, *World Rev. Nutr. Diet.,* Karger, Basel, 1990.
2. Simopoulos, A.P., Herbert, V., and Jacobson, B., *The Healing Diet: How to Reduce Your Risks and Live a Longer and Healthier Life If You Have a Family History of Cancer, Heart Disease, Hypertension, Diabetes, Alcoholism, Obesity, Food Allergies,* Macmillan Publishers, New York, 1995.
3. Simopoulos, A.P. and Nestel, P.J., Eds., *Genetic Variation and Dietary Response,* vol. 80, *World Rev. Nutr. Diet.,* Karger, Basel, 1997.
4. Simopoulos, A.P. and Pavlou, K.N., Eds., *Nutrition and Fitness 1: Diet, Genes, Physical Activity and Health,* vol. 89, *World Rev. Nutr. Diet.,* Karger, Basel, 2001.
5. Simopoulos, A.P., Genetic variation and dietary response: nutrigenetics/nutrigenomics, *Asian Pacific J. Clin. Nutr.,* 11(S6), S117, 2002.
6. Simopoulos, A.P. and Ordovas J., Eds., *Nutrigenetics and Nutrigenomics,* vol. 93, *World Rev. Nutr. Diet.,* Karger, Basel, 2004.
7. Eaton, S.B. and Konner, M., Paleolithic nutrition. A consideration of its nature and current implications. *New Engl. J. Med.,* 312, 283, 1985.
8. Simopoulos, A.P., Omega-3 fatty acids in health and disease and in growth and development, *Am. J. Clin. Nutr.,* 54, 438, 1991.
9. Simopoulos, A.P., Genetic variation and evolutionary aspects of diet, in *Antioxidants in Nutrition and Health,* Papas, A., Ed., CRC Press, Boca Raton, FL, 65, 1999.
10. Simopoulos, A.P., Evolutionary aspects of omega-3 fatty acids in the food supply, *Prostaglandins, Leukotrienes Essential Fatty Acids,* 60(5&6), 421, 1999.
11. Simopoulos, A.P., New products from the agri-food industry: the return of n-3 fatty acids into the food supply, *Lipids* 34(suppl), S297, 1999.
12. Cordain, L., Cereal grains: humanity's double-edged sword, *World Rev. Nutr. Diet.,* 84, 19, 1999.
13. Simopoulos, A.P., Ed., *Plants in Human Nutrition,* vol. 77, *World Rev. Nutr. Diet.,* Karger, Basel, 1995.
14. Zeghichi, S., Kallithraka, S., and Simopoulos, A.P., Nutritional composition of molokhia (Corchorus olitorius) and stamnagathi (Cicorium spinosum), *World Rev. Nutr. Diet.,* 91, 1, 2003.
15. Zeghichi, S., et al., The nutritional composition of selected wild plants in the diet of Crete, *World Rev. Nutr. Diet.,* 91, 22, 2003.
16. Simopoulos, A.P., Omega-3 fatty acids and antioxidants in edible wild plants, *Biol. Research,* 37(2), 263, 2004.
17. Simopoulos, A.P., Ed., *Evolutionary Aspects of Nutrition and Health. Diet, Exercise, Genetics and Chronic Disease,* vol. 84, *World Rev Nutr Diet.,* Karger, Basel, 1999.
18. Fanaian, M., et al., The effect of modified fat diet on insulin resistance and metabolic parameters in type II diabetes, *Diabetologia,* 39(suppl 1), A7, 1996.
19. Simopoulos, A.P., Is insulin resistance influenced by dietary linoleic acid and trans fatty acids? *Free Rad. Biol. & Med.,* 17(4), 367, 1994.
20. Simopoulos, A.P., Fatty acid composition of skeletal muscle membrane phospholipids, insulin resistance and obesity, *Nutr. Today,* 2, 12, 1994.

21. Simopoulos, A.P. and Robinson, J., *The Omega Diet. The Lifesaving Nutritional Program Based on the Diet of the Island of Crete*. HarperCollins, New York, 1999.
22. Eaton, S.B., Konner, M., and Shostak, M., Stone agers in the fast lane: chronic degenerative diseases in evolutionary perspective, *Am. J. Med.*, 84, 739, 1988.
23. Eaton, S.B., et al., Dietary intake of long-chain polyunsaturated fatty acids during the Paleolithic, *World Rev. Nutr. Diet.*, 83, 12, 1998.
24. Leaf, A. and Weber, P.C., A new era for science in nutrition, *Am. J. Clin. Nutr.*, 45, 1048, 1987.
25. Simopoulos, A.P., The Mediterranean diet: Greek column rather than an Egyptian pyramid, *Nutr. Today*, 30(2), 54, 1995.
26. Simopoulos, A.P., Overview of evolutionary aspects of w3 fatty acids in the diet, *World Rev. Nutr. Diet.*, 83, 1, 1998.
27. Simopoulos, A.P., Evolutionary aspects of diet and essential fatty acids, *World Rev. Nutr. Diet.*, 88, 18, 2001.
28. Simopoulos, A.P., Evolutionary aspects of diet: Fatty acids, insulin resistance and obesity, in *Obesity: New Directions in Assessment and Management*, VanItallie, T.B. and Simopoulos, A.P., Eds., Charles Press, Philadelphia, 1995, 241.
29. Simopoulos, A.P., Trans fatty acids, *Handbook of Lipids in Human Nutrition*, Spiller, G.A., Ed., CRC Press, Boca Raton, FL, 1995, 91.
30. Hunter, J.E., Omega-3 fatty acids from vegetable oils, in *Biological Effects and Nutritional Essentiality. Series A: Life Sciences*, vol. 171, Galli, C. and Simopoulos, A.P., Eds., Plenum Press, New York, 1989, 43.
31. Litin, L. and Sacks, F., Trans-fatty-acid content of common foods, *N. Engl. J. Med.*, 329(26), 1969.
32. Raper, N.R., Cronin, F.J., and Exler, J., Omega-3 fatty acid content of the US food supply, *J. Am. College Nutr.* 11(3), 304, 1992.
33. Crawford, M.A., Fatty acid ratios in free-living and domestic animals, *Lancet*, i, 1329, 1968.
34. Crawford, M.A., Gale, M.M., and Woodford, M.H., Linoleic acid and linolenic acid elongation products in muscle tissue of Syncerus caffer and other ruminant species, *Biochem. J.*, 115, 25, 1969.
35. Ledger, H.P., Body composition as a basis for a comparative study of some East African animals, *Symp. Zool. Soc. London*, 21, 289, 1968.
36. Cordain, L., et al., The fatty acid composition of muscle, brain, marrow and adipose tissue in elk: evolutionary implications for human dietary requirements, *World Rev. Nutr. Diet.*, 83, 225, 1998.
37. Sinclair, A.J., Slattery, W.J., and O'Dea, The analysis of polyunsaturated fatty acids in meat by capillary gas-liquid chromotography, *J. Food Sci. Agric.*, 33, 771, 1982.
38. Sugano, M. and Hirahara, F., Polyunsaturated fatty acids in the food chain in Japan, *Am. J. Clin. Nutr.*, 71(suppl), 189S, 2000.
39. Pella, D., et al., Effects of an Indo-Mediterranean diet on the omega-6/omega-3 ratio in patients at high risk of coronary artery disease: The Indian paradox, *World Rev. Nutr. Diet.*, 92, 74, 2003.
40. Sanders, T.A.B., Polyunsaturated fatty acids in the food chain in Europe, *Am. J. Clin. Nutr.*, 71(suppl), S176, 2000.
41. Brox, J.H., et al., Effects of cod liver oil on platelets and coagulation in familial hypercholesterolemia (type IIa), *Acta Med. Scand.*, 213, 137, 1983.
42. Joist, J.H., Baker, R.K., and Schonfeld, G., Increased in vivo and in vitro platelet function in type II- and type IV-hyperlipoproteinemia, *Thromb. Res.*, 15, 95, 1979.

43. Weber, P.C., Are we what we eat? Fatty acids in nutrition and in cell membranes: cell functions and disorders induced by dietary conditions, *Report No. 4*, Svanoy Foundation, Svanoybukt, Norway, 1989, 9.

44. Saynor, R., Verel. D., and Gillott, T., The long term effect of dietary supplementation with fish lipid concentrate on serum lipids, bleeding time, platelets and angina, *Atherosclerosis*, 50, 3, 1984.

45. Dehmer, G.J., et al., Reduction in the rate of early restenosis after coronary angioplasty by a diet supplemented with n-3 fatty acids, *N. Engl. J. Med.*, 319, 733, 1988.

46. Bottiger, L.E., Dyerberg, J., and Nordoy, A., Eds., N-3 Fish oils in clinical medicine. *J. Intern. Med*, 225(suppl 1), 1, 1989.

47. Cartwright, I.J., et al., The effects of dietary w-3 polyunsaturated fatty acids on erythrocyte membrane phospholipids, erythrocyte deformability, and blood viscosity in healthy volunteers, *Atherosclerosis*, 55, 267, 1985.

48. Lewis, R.A., Lee, T.H., and Austen, K.F., Effects of omega-3 fatty acids on the generation of products of the 5-lipoxygenase pathway, in *Health Effects of Polyunsaturated Fatty Acids in Seafoods*, Simopoulos, A.P., Kifer, R.R., and Martin, R.E., Eds., Academic Press, Orlando, 1986, 227.

49. Weber, P.C. and Leaf, A., Cardiovascular effects of w3 fatty acids: atherosclerotic risk factor modification by w3 fatty acids, in *Health Effects of w3 Polyunsaturated Fatty Acids in Seafoods*, Simopoulos, A.P., et al., Eds., vol. 66, *World Rev. Nutr. Diet.*, Karger, Basel, 1991, 218.

50. Barcelli, U.O., Glass-Greenwalt, P., and Pollak,V.E., Enhancing effect of dietary supplementation with omega-3 fatty acids on plasma fibrinolysis in normal subjects, *Thromb. Res.*, 39, 307, 1985.

51. Radack, K., Deck, C., and Huster, G., Dietary supplementation with low-dose fish oils lowers fibrinogen levels: a randomized, double-blind controlled study, *Ann. Intern. Med.*, 111, 757, 1989.

52. Sanders, T.A.B., Vickers, M., and Haines, A.P., Effect on blood lipids and hemostasis of a supplement of cod-liver oil, rich in eicosapentaenoic and docosahexaenoic acids, in healthy young men, *Clin. Sci.*, 61, 317, 1981.

53. Brown, A.J. and Roberts, D.C.K., Fish and fish oil intake: effect on hematological variables related to cardiovascular disease, *Thromb. Res.*, 64, 169, 1991.

54. De Caterina, R., et al., Vascular prostcyclin is increased in patients ingesting n-3 polyunsaturated fatty acids prior to coronary artery bypass surgery, *Circulation*, 82, 428, 1990.

55. Fox, P.L. and Dicorleto, P.E., Fish oils inhibit endothelial cell production of a platelet-derived growth factor-like protein, *Science*, 241, 453, 1988.

56. Shimokawa, H. and Vanhoutte, P.M., Dietary cod-liver oil improves endothelium dependent responses in hypercholesterolemic and atherosclerotic porcine coronary arteries, *Circulation*, 78, 1421, 1988.

57. Kremer, J.M., Jubiz, W., and Michalek, A., Fish-oil fatty acid supplementation in active rheumatoid arthritis, *Ann. Intern. Med.*, 106, 497, 1987.

58. Lee, T.H., et al., Effect of dietary enrichment with eicosapentaenoic and docosahexaenoic acids on in vitro neutrophil and monocyte leukotriene generation and neutrophil function, *N. Engl. J. Med.*, 312, 1217, 1985.

59. Endres, S., et al., The effect of dietary supplementation with n-3 polyunsaturated fatty acids on the synthesis of interleukin-1 and tumor necrosis factor by mononuclear cells, *N. Engl. J. Med.*, 320, 265, 1989.

60. Khalfoun, B., et al., Docosahexaenoic and eisosapentaenoic acids inhibit in vitro human endothelial cell production of interleukin-6, in *Eicosanoids and Other Bioactive Lipids in Cancer, Inflammation, and Radiation Injury 2*, Honn, K.V., et al., Eds, Plenum Press, New York, 1997.

61. Kremer, J.M., Lawrence, D.A., and Jubiz,W., Different doses of fish-oil fatty acid ingestion in active rheumatoid arthritis: a prospective study of clinical and immunological parameters, in *Dietary w3 and w6 Fatty Acids: Biological Effects and Nutritional Essentiality*, Galli, C. and Simopoulos, A.P., Eds., Plenum Publishing, New York, 1989, 343.

62. Robinson, D.R. and Kremer, J.M., Summary of Panel G: rheumatoid arthritis and inflammatory mediators, *World Rev. Nutr. Diet.*, 66, 44, 1991.

63. Ross, R., Atherosclerosis — an inflammatory disease, *N. Engl. J. Med.*, 340, 115, 1999.

64. Springer, T.A., Traffic signals for lymphocyte recirculation and leukocyte emigration: the multistep paradigm, *Cell*, 76, 301, 1994.

65. Davies, M.J., et al., The expression of the adhesion molecules ICAM-1, VCAM-1, PECAM, and E-selectin in human atherosclerosis, *J. Pathol.*, 171, 223, 1993.

66. O'Brien, K.D., et al., Vascular cell adhesion molecule-1 is expressed in human coronary atherosclerotic plaques: implications for the mode of progression of advanced coronary atherosclerosis, *J. Clin. Invest.* 92, 945, 1993.

67. Poston, R.N., et al., Expression of intercellular adhesion molecule-1 in atherosclerotic plaques, *Am. J. Pathol.* 140, 665, 1992.

68. Richardson, M., et al., Increased expression in vivo of VCAM-1 and E-selectin by the aortic endothelium of normolipemic and hyperlipemic diabetic rabbits, *Arterioscler. Thromb.* 14, 760, 1994.

69. Simopoulos, A.P., The role of fatty acids in gene expression: health implications, *Ann. Nutr. Metab.*, 40, 303, 1996.

70. Ge, Y-L., et al., Effects of adenoviral transfer of Caenorhabditis elegans n-3 fatty acid desaturase on the lipid profile and growth of human breast cancer cells, *Anticancer Research*, 22, 537, 2002.

71. Kang, Z.B., et al., Adenoviral transfer of Caenorhabditis elegans n-3 fatty acid desaturase optimizes fatty acid composition in mammalian heart cells, *Proc. Natl. Acad. Sci. U.S.A.*, 98, 4050, 2001.

72. Kang, J.X., et al., *Fat-1* mice convert n-6 to n-3 fatty acids, *Nature*, 427, 504, 2004.

73. de Lorgeril, M., et al., Mediterranean alpha-linolenic acid-rich diet in secondary prevention of coronary heart disease, *Lancet*, 343, 1454, 1994.

74. de Lorgeril, M. and Salen, P., Modified Cretan Mediterranean diet in the prevention of coronary heart disease and cancer, *World Rev. Nutr. Diet.*, 87, 1, 2000.

75. Renaud, S., et al., Cretan Mediterranean diet for prevention of coronary heart disease. *Am. J. Clin. Nutr.*, 61(suppl), 1360S, 1995.

76. Cleland, L.G., et al., Linoleate inhibits EPA incorporation from dietary fish-oil supplements in human subjects, *Am. J. Clin. Nutr.*, 55, 395, 1992.

77. GISSI-Prevenzione Investigators, Dietary supplementation with n-3 polyunsaturated fatty acids and vitamin E after myocardial infarction: results of the GISSI-Prevenzione trial, *Lancet*, 354, 447, 1999.

78. Li, Y., Kang, J.X., and Leaf, A., Differential effects of various eicosanoids on the contraction of cultured neonatal rat cardiomyocytes, *Prostaglandins*, 54, 511, 1997.

79. Burr, M.L., et al., Effect of changes in fat, fish and fibre intakes on death and myocardial reinfarction: diet and reinfarction trial (DART), *Lancet*, 2, 757, 1989.

80. Singh, R.B. et al., Randomized controlled trial of cardioprotective diet in patients with recent acute myocardial infarction: results of one year follow up, *Brit. Med. J.*, 304, 1015, 1992.

81. Singh, R.B., et al., Randomized, double-blind, placebo-controlled trial of fish oil and mustard oil in patients with suspected acute myocardial infarction: the Indian experiment of infarct survival — 4, *Cardiovasc. Drugs Ther.*, 11, 485, 1997.

82. Raheja, B.S., et al., Significance of the n-6/n-3 ratio for insulin action in diabetes, *Ann. N.Y. Acad. Sci.*, 683, 258, 1993.

83. James, M.J. and Cleland, L.G., Dietary n-3 fatty acids and therapy for rheumatoid arthritis, *Semin. Arthritis Rheum.*, 27, 85, 1997.

84. Broughton, K.S., et al. Reduced asthma symptoms with n-3 fatty acid ingestion are related to 5-series leukotriene production, *Am. J. Clin. Nutr.*, 65(4), 1011, 1997.

85. Laitinen, L.A., Laitinen, A., and Haahtela, T., Airway mucosal inflammation even in patients with newly diagnosed asthma, *Am. Rev. Respir. Dis.*, 147(3), 697, 1993.

86. Bartram, H.-P., et al., Effects of fish oil on rectal cell proliferation, mucosal fatty acids, and prostaglandin E_2 release in healthy subjects, *Gastroenterology*, 105, 1317, 1993.

87. Bartram, H.-P., et al. Missing anti-proliferative effect of fish oil on rectal epithelium in healthy volunteers consuming a high-fat diet: potential role of the n-3:n-6 fatty acid ratio, *Eur. J. Cancer Prev.*, 4, 231, 1995.

88. Maillard, V., et al., N-3 and n-6 fatty acids in breast adipose tissue and relative risk of breast cancer in a case-control study in Tours, France, *Int. J. Cancer*, 98, 78, 2002.

89. Maes, M., et al., Fatty acid composition in major depression: decreased omega 3 fractions in cholesteryl esters and increased C20:4 omega 6/C20:5 omega 3 ratio in cholesteryl esters and phospholipids, *J. Affect Disord.*, 38(1), 35, 1996.

90. Maes, M., et al., Lower serum high-density lipoprotein cholesterol (HDL-C) in major depression and in depressed men with serious suicidal attempts: relationship with immune-inflammatory markers, *Acta Psychiatr. Scand.*, 95(3), 212, 1997.

91. Peet, M., et al., Depletion of omega-3 fatty acid levels in red blood cell membranes of depressive patients, *Biol. Psychiatry*, 43(5), 315, 1998.

92. Stoll, A.L., et al., Omega-3 fatty acids in bipolar disorder: a preliminary double-blind placebo-controlled trial, *Arch. Gen. Psychiatry* 56(5), 407, 1999.

93. Locke, C.A. and Stoll, A.L., Omega-3 fatty acids in major depression, *World Rev. Nutr. Diet.*, 89, 173, 2001.

94. Simopoulos, A.P. and Cleland, L.G., Eds., *Omega-6/Omega-3 Essential Fatty Acid Ratio: The Scientific Evidence*, vol. 92, *World Rev. Nutr. Diet.,* Karger, Basel, 2003.

95. Simopoulos, A.P., N-3 fatty acids and human health: defining strategies for public policy, *Lipids*, 36, S83, 2001.

96. Graber, R., Sumida, C., and Nunez, E.A., Fatty acids and cell signal transduction, *J. Lipid Mediat. Cell Signal*, 9, 91, 1994.

97. Clarke, S.D. and Jump, D.B., Dietary polyunsaturated fatty acid regulation of gene transcription, *Annu. Rev. Nutr.*, 14, 83, 1994.

98. Clarke, S.D., Romsos, D.R., and Leveille, G.A., Differential effects of dietary methylesters of long chain saturated and polyunsaturated fatty acids on rat liver and adipose tissue lipogenesis, *J. Nutr.*, 107, 1170, 1977.

99. Clarke, S.D., Armstrong, M.K., and Jump, D.B., Nutritional control of rat liver fatty acid synthase and S14 mRNA abundance, *J. Nutr.*, 120, 218, 1990.

100. Clarke, S.D. and Jump, D.B., Fatty acid regulation of gene expression: a unique role for polyunsaturated fats, in *Nutrition and Gene Expression*, Berdanier, C. and Hargrove, J.L., Eds., CRC Press, Boca Raton, FL, 1993, 227.

101. Clarke, S.D. and Jump, D.B., Polyunsaturated fatty acid regulation of hepatic gene transcription, *Lipids*, 31(suppl), 7, 1996.

102. Jump, D.B., et al., Coordinate regulation of glycolytic and lipogenic gene expression by polyunsaturated fatty acids, *J. Lipid Res.*, 35, 1076, 1994.

103. Tebbey, P.W., et al., Arachidonic acid down-regulates the insulin-dependent glucose transporter gene (Glut 4) in 3T3-L1 adipocytes by inhibiting transcription and enhancing mRNA turnover, *J. Biol. Chem.*, 269, 639, 1994.

104. Limatta, M., et al., Dietary PUFA interfere with the insulin glucose activation of L-type pyruvate kinase. *Mol. Endocrinol.*, 8, 1147, 1994.

105. Ntambi, J.M., Dietary regulation of stearoyl-CoA desaturase I gene expression in mouse liver, *J. Biol. Chem.*, 267, 10925, 1991.

106. DeWillie, J.W. and Farmer, S.J., Linoleic acid controls neonatal tissue-specific stearoyl-CoA desaturase mRNA levels, *Biochim. Biophys. Acta.*, 1170, 291, 1993.

107. Sellmayer, A., Danesch, U., and Weber, P.C., Effects of different polyunsaturated fatty acids on growth-related early gene expression and cell growth, *Lipids*, 31, S37, 1996.

108. De Caterina, R. and Libby, P., Control of endothelial leukocyte adhesion molecules by fatty acids, *Lipids*, 31(suppl), 57, 1996.

109. Robinson, D.R., et al., Dietary marine lipids suppress the continuous expression of interleukin-1B gene transcription, *Lipids*, 31(suppl):23, 1996.

110. Kaminski, W.E., et al., Dietary omega-3 fatty acids lower levels of platelet-derived growth factor mRNA in human mononuclear cells, *Blood*, 81, 1871, 1993.

111. Holian, O. and Nelson, R., Action of long-chain fatty acids on protein kinase C activity: comparison of omega-6 and omega-3 fatty acids, *Anticancer Res.*, 12, 975, 1992.

112. Lucas, A., et al., Breast milk and subsequent intelligence quotient in children born premature, *Lancet*, 339, 261, 1992.

113. Eurocat Working Group, Prevalence of neural tube defects in 20 regions of Europe and the impact of prenatal diagnosis 1980–86. *J. Epidemiol. Community Health*, 45, 52, 1991.

114. Dwyer, J.H., et al., Arachidonate 5-lipoxygenase promoter genotype, dietary arachidonic acid, and atherosclerosis, *N. Engl. J. Med.*, 350, 29, 2004.

CHAPTER 11

Beyond Fiber: Whole Grains and Health

Joanne Slavin

CONTENTS

HISTORICAL BACKGROUND ON WHOLE GRAINS

Whole grains became part of the human diet with the advent of agriculture about 10,000 years ago.[1] For the last 3000 to 4000 years, a majority of the world's population has relied upon whole grains as a main portion of the diet. In North America, wheat, oats, barley, and rye were harvested as staple foods as early as the American Revolution. It is only within the past 100 years that a majority of the population has consumed refined grain products. Prior to this time, gristmills were used for grinding grains. They did not completely separate the bran and germ from the white endosperm and produced limited amounts of purified flour. In 1873, the roller mill was introduced and it more efficiently separated the bran and germ from the endosperm. Widespread use of the roller mill fueled an increasing consumer demand for refined grain products and was a significant factor in the dramatic decline in whole-grain consumption observed from about 1870 to 1970.[1]

Health aspects of whole grains have long been known. In the 4th century BC, Hippocrates, the father of medicine, recognized the health benefits of whole-grain bread. More recently, physicians and scientists in the early 1800s to mid 1900s recommended whole grains to prevent constipation. The "fiber hypothesis," published in the early 1970s, suggested that whole foods, such as whole grains, fruits, and vegetables, provide fiber along with other constituents that have health benefits.

WHAT ARE WHOLE GRAINS?

The major cereal grains include wheat, rice, and corn, with oats, rye, barley, triticale, sorghum, and millet as minor grains. In the U.S., the most commonly consumed grains are wheat, oats, rice, corn, and rye, with wheat constituting 66 to 75% of the total. Buckwheat, wild rice, and amaranth are not botanically true grains but are typically associated with the grain family due to their similar composition. All grains have a barklike, protective hull, beneath which are the endosperm, bran, and germ. The germ contains the plant embryo. The endosperm supplies food for the growing seedling. Surrounding the germ and the endosperm is the outer covering, or bran, which protects the grain from its environment, including weather, insects, molds, and bacteria.

About 50 to 75% of the endosperm is starch, and it is the major energy supply for the embryo during germination of the kernel. The endosperm also contains storage proteins, typically 8 to 18%, along with cell wall polymers. Relatively few vitamins, minerals, fiber, or phytochemicals are located in the endosperm fraction. The germ is a relatively minor contributor to the dry weight of most grains (typically 4 to 5% in wheat and barley). The germ of corn contributes a much higher proportion to the total grain structure than that of wheat, barley, or oats.

WHO CONSUMES WHOLE GRAINS?

Increased whole-grain consumption is widely promoted. Grain products comprise the base of the U.S. Department of Agriculture (USDA) Food Guide Pyramid, which suggests that several of the recommended 6 to 11 servings of grain products per day should be from whole grains. The 2005 Dietary Guidelines for Americans placed particular emphasis on eating more whole-grain foods. It is recommended that at least three servings, or one-half of grain foods consumed daily, be whole grain.

Americans consume far less than the recommended three servings of whole grains on a daily basis. According to a survey of Americans 20 years and older,[2] total grain intake was 6.7 servings per day with only 1.0 of these servings being whole grain. Only 8% of the study participants consumed the recommended three servings of whole grains on a daily basis. Another investigation[3] reported that less than 2% of the study population consumed two or more whole grain servings per day, and 23% consumed no whole grains over the two-week reporting period.

Average whole-grain consumption for this study was 0.5 servings per day. A study of U.S. children and teens reported consumption of whole grains was less than one serving per day.[4]

Whole-grain intake studies in other countries find similar results. Except for parts of Scandinavia where whole-grain breads are the norm, whole-grain consumption is low.[5] In the U.K., median consumption of whole grains was less than one serving per day.[5] Efforts have been made to increase whole-grain consumption. A whole-grain health claim has also been approved in the U.S. For a whole-grain food to meet the whole-grain health claim standards, the food must include 51% whole-grain flour by weight of final product and must contain 1.7 grams of dietary fiber.

COMPONENTS IN WHOLE GRAINS

The bran and germ fractions derived from conventional milling provide a majority of the biologically active compounds found in a grain. Specific nutrients include high concentrations of B vitamins (thiamin, niacin, riboflavin, and pantothenic acid) and minerals (calcium, magnesium, potassium, phosphorus, sodium, and iron), elevated levels of basic amino acids (e.g., arginine and lysine), and elevated tocol levels in the lipids. Numerous phytochemicals, some common in many plant foods (phytates and phenolic compounds) and some unique to grain products (avenanthramides, avenalumic acid), are responsible for the high antioxidant activity of whole-grain foods.[6]

In developed countries, such as the U.S. and Europe, grains are generally subjected to some type of processing, milling, heat extraction, cooking, parboiling, or other technique prior to consumption. Commercial cereals are usually extruded, puffed, flaked, or otherwise altered to make a desirable product. Most research finds that processing of whole grains does not remove biologically important compounds.[7] Analysis of processed breads and cereals indicate that they are a rich source of antioxidants.[6] Processing may open up the food matrix, thereby allowing the release of tightly bound phytochemicals from the grain structure.[8] Studies with rye find that many of the bioactive compounds are stable during food processing, and their levels may even be increased with suitable processing.[9]

Components in whole grains associated with improved health status include lignans, tocotrienols, phenolic compounds, and antinutrients including phytic acid, tannins, and enzyme inhibitors. In the grain refining process the bran is removed, resulting in loss of dietary fiber, vitamins, minerals, lignans, phytoestrogens, phenolic compounds, and phytic acid. Thus, refined grains are more concentrated in starch since most of the bran and some of the germ are removed in the refining process.

Recent studies have suggested that components of whole grains could be used as biomarkers for whole-grain intake. Alkylresorcinols are good markers of whole-grain wheat and rye in foods, and their analysis in biological samples may indicate whole-grain consumption.[10] Additionally, plasma enterolactone, a phytoestrogen, has been positively linked to whole-grain consumption in Danish women.[11]

WHOLE GRAINS AND CARDIOVASCULAR DISEASE

Cardiovascular disease (CVD) is the number one cause of death and disability of both men and women in this country. There is strong epidemiological and clinical evidence linking consumption of whole grains to a reduced risk for coronary heart disease.[12] Morris et al.[13] followed 337 subjects for 10 to 20 years and concluded that a reduction in heart disease risk was attributable to a higher intake of cereal fiber. Brown et al.[14] concluded that soluble fiber from different fiber sources was associated with small but significant decreases in total cholesterol. Other compounds in grains, including antioxidants, phytic acid, lectins, phenolic compounds, amylase inhibitors, and saponins have all been shown to alter risk factors for coronoary heart disease (CHD). It is likely that the combination of compounds in grains, rather than any one component, explains its protective effects in CHD.

Large prospective epidemiologic studies have found a moderately strong association between whole-grain intake and decreased CHD risk. Postmenopausal women (34,492), aged 55 to 69 years and free of CHD were followed in the large prospective Iowa Women's Health Study for occurrence of CHD mortality (n = 387) between baseline (1986) and 1994.[15] Whole-grain intake was determined by 7 items in a 127-item food frequency questionnaire that was used to divide participants into quintiles based on mean servings of whole-grain intake per day. The risk reduction in higher whole-grain intake quintiles was controlled for more than 15 confounding variables and was not explained by adjustment for dietary fiber intake. This suggests that whole-grain components other than dietary fiber may reduce risk for CHD.

In a Finnish study, 21,930 male smokers (aged 50 to 69 years) were followed for 6.1 years.[16] Reduced risk of CHD death was associated with increased intake of rye products. Rimm et al.[17] examined the association between cereal intake and risk for myocardial infarction (MI) in 43,757 U.S. health professionals, aged 40 to 75 years. Cereal fiber was most strongly associated with reduced risk for MI with a 0.71 decrease in risk for each 10 g increase in cereal fiber intake.

The Nurses' Health Study, a large, prospective cohort study of U.S. women followed up for 10 years, was also used to examine the relationship between grain intake and cardiovascular risk.[18] A total of 68,782 women aged 37 to 64 years without previously diagnosed angina, myocardial infarction, stroke, cancer, hypercholesterolemia, or diabetes at baseline were studied. Dietary data were collected with a validated semiquantitative food frequency questionnaire. After controlling for age, cardiovascular risk factors, dietary factors, and multivitamin supplement use, the relative risk was 0.77 (95% CI, 0.57 to 1.04). For a 10 g/day increase in total fiber intake (the difference between the lowest and highest quintiles), the multivariate RR of total CHD events was 0.81 (95% CI, 0.66 to 0.99). Among different sources of dietary fiber (cereal, vegetable, fruit), only cereal fiber was strongly associated with a reduce risk of CHD (multivariate RR, 0.63; 95% CI, 0.49 to 0.81 for each 5 g/d increase in cereal fiber). The authors conclude that higher fiber intake, particularly from cereal sources, reduces the risk of CHD.

Feeding studies have looked at different biomarkers relevant to cardiovascular disease. Truswell[19] concluded that enough evidence exists that whole-grain products may reduce the risk of coronary heart disease. Katz et al.[20] measured the effect of oat and wheat cereals on endothelial responses in human subjects. They report that month-long, daily supplementation with either whole-grain oat or wheat cereal may prevent postprandial impairment of vascular reactivity in response to a high-fat meal. In a randomized controlled clinical trial, consumption of whole-grain and legume powder reduced insulin demand, lipid peroxidation, and plasma homocysteine concentrations in patients with coronary artery disease.[21] Finally, consumption of whole-grain oat cereal was associated with improved blood pressure control and reduced the need for antihypertensive medications.[22] Thus, clinical studies to date support that whole-grain consumption can improve biomarkers relevant to diabetes and cardiovascular disease.

Food consumption patterns that include whole grains also appear protective for cardiovascular disease. Van Dam et al.[23] report that intake of refined diets that do not include whole grains were associated with higher serum cholesterol levels and lower intakes of micronutrients. A prudent dietary pattern, including intake of whole grains, was associated with lower C-reactive protein levels and endothelial dysfunction, an early step in the development of atherosclerosis.[24] Whole-grain food intake was also associated with lower levels of C-reactive protein in the Nurses' Health Study.[25]

Since whole grains are the predominant dietary fiber source in the U.S., it is difficult to separate the protection of dietary fiber from whole grains. Jensen et al.[26] examined intakes of whole grains, bran, and germ and risk of coronary heart disease from food frequency data in the Health Professionals Follow-Up Study. Added germ was not associated with CHD risk, and the authors conclude that the study supports the reported beneficial association of whole-grain intake with CHD and suggests that the bran component of whole grains could be a key factor in this relation.

WHOLE GRAINS AND BLOOD GLUCOSE

It is well accepted that glucose and insulin are linked to chronic diseases, especially diabetes. Whole-grain consumption is part of a healthy diet described as the "prudent" diet. Epidemiologic studies consistently show that risk for type 2 diabetes mellitus is decreased with consumption of whole grains.[27,28] Whole grains are now recommended by the American Diabetes Association for diabetes prevention.[29]

Whole foods are also known to slow digestion and absorption of carbohydrates. Postprandial blood glucose and insulin responses are greatly affected by food structure. Any process that disrupts the physical or botanical structure of food ingredients will increase the plasma glucose and insulin responses. Food structure was found to be more important than gelatinization or presence of viscous dietary fiber in determining glycemic response.[30] Another study found the importance of preserved structure in foods as an important determinant of glycemic response in diabetics.[31] Refining grains tends to increase glycemic response and, thus, whole grains should slow glycemic response.[32]

Intact whole grains of barley, rice, rye, oats, corn, buckwheat, and wheat have glycemic indices (GI) of 36 to 81, with barley and oats having the lowest values.[33] Lower blood glucose levels and decreased insulin secretion have been seen in both normal and diabetic subjects while consuming a low GI (=67) diet containing pumpernickel bread with intact whole grains, bulgur (parboiled wheat), pasta, and legumes compared to a high GI (=90) diet containing white bread and potatoes.

Heaton et al.[34] compared glucose response when subjects consumed whole grains, cracked grains, whole-grain flour, and refined grain flour. Plasma insulin responses increased stepwise, with whole grains less than cracked grains less than coarse flour less than fine flour. Oat-based meals evoked smaller glucose and insulin responses than wheat- or maize-based meals. Particle size influenced the digestion rate and consequent metabolic effects of wheat and maize, but not oats. The authors suggest the increased insulin response to finely ground flour may be relevant to the etiology of diseases associated with hyperinsulinemia and to the management of diabetes.

Some feeding studies have been conducted to evaluate the relationship between whole grains and glucose metabolism. Pereiera et al.[35] tested the hypothesis that whole-grain consumption improves insulin sensitivity in overweight and obese adults. Eleven overweight or obese hyperinsulinemic adults aged 25 to 56 years consumed 2 diets, each for 6 weeks. Diets were identical, except that whole-grain products replaced refined grain products. At the end of each treatment, subjects consumed 355 ml of a liquid mixed meal, and blood samples were taken over 2 hours. Fasting insulin was 10% lower during consumption of the whole-grain diet. The authors conclude that insulin sensitivity may be an important mechanism whereby whole-grain foods reduce the risk of type 2 diabetes and heart disease.

Juntunen et al.[36] evaluated what factors in grain products affected human glucose and insulin responses. They fed the following grain products: whole-kernel rye bread, whole-meal rye bread containing oat beta-glucan concentrate, dark durum wheat pasta, and wheat bread made from white wheat flour. Glucose responses and the rate of gastric emptying after consumption of the two rye breads and pasta did not differ from those after consumption of white wheat bread. Insulin, glucose-dependent insulinotropic polypeptide, and glucagon-like peptide 1, were lower after consumption of rye breads and pasta than after consumption of white wheat bread. These results support that postprandial insulin responses to grain products are determined by the form of food and botanical structure rather than by the amount of fiber or the type of cereal in the food.

Intake of fiber from whole-grain cereals has also been found to be inversely related to type 2 diabetes. In a long-term study of almost 90,000 women[37] and in a similar study of about 45,000 men,[38] researchers found that those with higher intakes of cereal fiber had about a 30% lower risk for developing type 2 diabetes, compared to those with the lowest intakes. Additionally, the Iowa Women's Health Study found that dietary fiber and whole-grain intake were protective against type 2 diabetes.[39] In another study,[40] individuals consuming large amounts of refined grains and small amounts of whole grain had a 57% higher risk of type 2 diabetes than did those consuming large amounts of whole grains. In the Health Professionals Follow-Up Study, an investigation following 42,898 men, a 37% lower risk of type 2 diabetes was associated with about three servings/day of whole-grain intake.[41]

Whole grains are good sources of dietary magnesium, fiber, and vitamin E, which are involved in insulin metabolism. Relatively high intakes of these nutrients from whole grains may prevent hyperinsulinemia. Whole grains may also influence insulin levels through beneficial effects on satiety and body weight. However, even after adjusting for body mass index, studies have found a strong inverse relationship between whole-grain intake and fasting insulin levels.

Montonen et al.[42] reported an inverse association between whole-grain intake and risk of type 2 diabetes in a cohort study. Cereal fiber intake was also associated with a reduced risk of type 2 diabetes. Liu[43] pooled data from prospective cohort studies of whole-grain intakes and type 2 diabetes. The summary estimate of relative risk was 0.70. These studies cannot provide direct causal proof of the effects of whole grains in lowering the risk of type 2 diabetes, since confounding remains an alternative explanation in nonrandomized settings.

The synergistic effect of several whole-grain components, such as phytochemicals, vitamin E, magnesium, or others, may be involved in the reduction of risk for type 2 diabetes. McKeown et al.[44] reported that whole-grain intake was inversely associated with body mass index and fasting insulin in the Framingham Offspring Study. Juntunen et al.[45] fed high-fiber rye bread and white-wheat bread to postmenopausal women and measured glucose and insulin metabolism. Acute insulin response increased significantly more during the rye bread periods than during the wheat bread period. They suggest that high-fiber rye bread appears to enhance insulin secretion, possibly indicating improvement of beta cell function. Pereira et al.[35] did find improvements in insulin sensitivity with whole-grain consumption.

WHOLE GRAINS AND CANCER

There is substantial scientific evidence that whole grains as commonly consumed reduce the risk of cancer. In a meta-analysis of whole-grain intake and cancer, whole grains were found to be protective in 46 of 51 mentions of whole-grain intake, and in 43 of 45 mentions after exclusion of 6 mentions with design/reporting flaws or low intake.[46] Odds ratios were <1 in 9 of 10 mentions of studies of colorectal cancers and polyps, 7 of 7 mentions of gastric and 6 of 6 mentions of other digestive tract cancers, 7 of 7 mentions of hormone-related cancers, 4 of 4 mentions of pancreatic cancer, and 10 of 11 mentions of 8 other cancers. The pooled odds ratio was similar in studies that adjusted for few or many covariates. A systemic review of case-control studies conducted using a common protocol in Northern Italy between 1983 and 1996 indicates that a higher frequency of whole-grain consumption is associated with reduced risk for cancer.[47] Whole grains were consumed primarily as whole-grain bread and some whole-grain pasta in the Italian studies.

Other studies have demonstrated a lower risk for specific cancers, such as stomach,[48] mouth/throat and upper digestive tract,[49] endometrial,[50] and epithelial ovarian cancer.[51] Epidemiological studies have reported that higher serum insulin levels are associated with increased risk of colon, breast, and possibly other cancers. Reduction of these insulin levels by whole grains may be an indirect way in which the reduction in cancer risk occurs.

Dietary factors, such as fiber, vitamin B6, and phytoestrogen intake, and lifestyle factors such as exercise, smoking, and alcohol use, which are controlled for in most epidemiologic studies, do not explain the apparent protective effect of whole grains against cancer, again suggesting that it is the whole-grain "package" that is effective. Several theories have been offered to explain the protective effects of whole grains. Because of the complex nature of whole grains, there are many potential mechanisms that could be responsible their protective properties.

Several mechanisms have been proposed for the protective action of the dietary fiber found in whole grains. Increased fecal bulk and decreased transit time allow less opportunity for fecal mutagens to interact with the intestinal epithelium. Secondary bile acids are thought to promote cell proliferation, thus allowing increased opportunity for mutations to occur and abnormal cells to multiply. The effect of fiber on the actions of bile acids may be attributable to the binding or diluting of bile acids.

Whole grains also contain several antinutrients, such as protease inhibitors, phytic acid, phenolics, and saponins, which until recently were thought to have only negative nutritional consequences. Some of these antinutrient compounds may act as cancer inhibitors by preventing the formation of carcinogens and by blocking the interaction of carcinogens with cells. Other potential mechanisms linking whole grains to reduced cancer risk include large bowel effects, antioxidants, alterations in blood glucose levels, weight loss, hormonal effects, and the influence of numerous biologically active compounds.

Hormonally active compounds in grains called lignans may protect against diseases. Lignans are compounds processing a 2,3-dibenzylbutane structure and exist as minor constituents of many plants where they form the building blocks for the formation of lignin in the plant cell wall. The plant lignans secoisolariciresinol and matairesinol are converted by human gut bacteria to the mammalian lignans, enterolactone and enterodiol. Due to the association of lignan excretion with fiber intake, it is assumed that plant lignans are contained in the outer layers of the grain. Concentrated sources of lignans include whole-grain wheat, whole-grain oats, and rye meal. Seeds are also concentrated sources of lignans, including flaxseeds (the most concentrated source), pumpkin seeds, caraway seeds, and sunflower seeds.

Grains and other high-fiber foods increase urinary lignan excretion, an indirect measure of lignan content in foods.[52] Differences in metabolism of phytoestrogens among individuals have been noted. Adlercreutz et al.[53] found total urinary lignan excretion in Finnish women to be positively correlated with total fiber intake, total fiber intake per kg body weight, and grain fiber intake per kg body weight. Similarly, the geometric mean excretion of enterolactone was positively correlated with the geometric mean intake of dietary grain products (kcal/day) of five groups of women (r = 0.996).

Due to the association of lignan excretion with fiber intake, plant lignans are probably concentrated in the outer layers of the grain. Because current processing techniques eliminate this fraction of the grain, lignans may not be found in processed grain products on the market and would only be found in whole-grain foods.

Serum enterolactone was measured in a cross-sectional study in Finnish adults.[54] In men, serum enterolactone concentrations were positively associated with consumption of whole-grain products. Variability in serum enterolactone concentration was great, suggesting the role of gut microflora in the metabolism of lignans may be important. Kilkkinen et al.[55] also report that intake of lignans is associated with serum enterolactone concentration in Finnish men and women. They suggest that serum enterolactone is a feasible biomarker of lignan intake.

Jacobs et al.[56] found similar results in a U.S. study. Subjects were fed either whole-grain or refined-grain foods for 6 weeks. Most of the increase in serum enterolactone when eating the whole-grain diet occurred within 2 weeks, though the serum enterolactone difference between whole-grain and refined-grain diets continued to increase throughout the 6-week study. Serum enterolactone was associated with reduced cardiovascular disease-related and all-cause death in middle-aged Finnish men.[57] The authors suggest that this evidence supports the importance of whole-grain foods, fruits, and vegetables in the prevention of premature death from CVD.

Few studies have looked at the direct effects of feeding defined whole-grain diets to humans. McIntosh et al.[58] fed rye and wheat foods to overweight middle-aged men and measured markers of bowel health. The men were fed low-fiber cereal grain foods providing 5 grams of dietary fiber for the refined-grain diet and 18 grams of dietary fiber for the whole-grain diet, either high in rye or wheat. This was in addition to a baseline diet that contained 14 grams of dietary fiber. Both the high-fiber rye and wheat foods increased fecal output by 33 to 36% and reduced fecal β-glucuronidase activity by 29%. Postprandial plasma insulin was decreased by 46 to 49% and postprandial plasma glucose was decreased by 16 to 19%. Rye foods were associated with significantly increased plasma enterolactone and fecal butyrate, relative to wheat and low-fiber diets. The authors conclude that rye appears more effective than wheat in overall improvement of biomarkers of bowel health.

BODY WEIGHT REGULATION

Preliminary studies suggest an association between whole-grain intake and the regulation of body weight. In the Coronary Artery Risk Development in Young Adults (CARDIA) Study, whole grains were inversely associated with BMI and waist-hip ratio at baseline and 7 years later.[59] Although the differences were modest, the risk for weight gain and the development of overweight or obesity could be substantially decreased if the associations are true. A 10-year follow-up to the CARDIA study looked at dietary fiber, of which whole grains are a good source. Individuals with the highest dietary fiber intake (> 21 g/2000 kcal) gained approximately 8 fewer pounds of weight than did those with the lowest intake (< 12 g/2000 kcal). Similar results were found for the waist-hip ratio.[60]

Several factors may explain the influence of whole grains on body weight regulation. The high-volume, low-energy density and the relatively lower palatability of whole-grain foods may promote satiation (regulation of energy intake per eating occasion through effects of hormones influenced by chewing and swallowing mechanics). Additionally, whole grains may enhance satiety (delayed return of hunger following a meal) for up to several hours following a meal. Grains rich in viscous soluble fibers (e.g., oats and barley) tend to increase intraluminal viscosity, prolong gastric emptying time, and slow nutrient absorption in the small intestine. Although preliminary evidence suggests whole grains may influence body weight regulation, additional epidemiological studies and clinical trials are needed. Newby et al.[61] report that a healthy eating pattern, including consumption of whole grains, is associated with smaller gains in BMI and waist circumference in the ongoing Baltimore Longitudinal Study of Aging.

Weight gain among men in the Health Professionals Follow-Up Study was followed over 8 years and compared to changes in whole-grain, bran, and cereal fiber intake.[62] The increased consumption of whole grains was inversely related to weight gain, and the associations persisted after changes in added bran or fiber intakes were accounted for. This suggests that the components in whole grains beyond dietary fiber may contribute to favorable metabolic changes that reduce long-term weight gain.

ALL-CAUSE MORTALITY

Several epidemiological studies suggest whole grains reduce the risk for all-cause mortality or all-cause death. In the Iowa Women's Health Study, whole grains and cereal fiber lowered all-cause death in postmenopausal women,[63,64] and a Norwegian study showed a lower mortality rate for men and women with a high whole-grain bread intake.[65] Liu et al.[66] reported that both total mortality and CVD-specific mortality were inversely associated with whole-grain but not refined-grain breakfast cereal intake in the Physicians' Health Study.

CONCLUSION

Whole grains are rich in many components, including dietary fiber, starch, fat, antioxidant nutrients, minerals, vitamin, lignans, and phenolic compounds that have been linked to reduced risk of coronary heart disease, cancer, diabetes, obesity, and other chronic diseases. Most of the protective components are found in the germ and bran, which are reduced in the grain refining process. Based on epidemiological studies and biologically plausible mechanisms, the scientific evidence shows that regular consumption of whole-grain foods provides health benefits in terms of reduced rates of CHD and several forms of cancer. It may also help regulate blood glucose levels. More research is needed on the mechanisms for this protection. Also, some components in whole grains may be most important in this protection and should be retained in food processing.

Dietary intake of whole grains falls short of current recommendations to eat at least 3 servings a day. The whole-grain health claim should increase consumption of whole-grain foods in the American population. This is in keeping with the Food Pyramid educational materials, which recommend that a minimum of six servings of grain foods be eaten each day, with at least three of those servings as whole grains. Successful implementation of these recommendations will require the cooperative efforts of industry, government, academia, nonprofit health organizations, and the media. Additional work is needed to confirm the health benefits of whole grains, develop processing techniques that will improve the palatability of whole-grain products, and educate consumers about the benefits of whole-grain consumption.

REFERENCES

1. Spiller, G.A., Whole grains, whole wheat, and white flours in history. In *Whole Grain Foods in Health and Disease,* Marquart, L., Slavin, J.L., and Fulcher, R.G., Eds., Eagan Press, St. Paul, MN, 2002.
2. Cleveland, L.E., et al., Dietary intake of whole grains. *J. Am. Coll. Nutr.* 18, 331S, 2000.
3. Albertson, A.M. and Tobelmann, R.C., Consumption of grain and whole-grain foods by an American population during the years 1990 to 1998. In *Whole Grain Foods in Health and Disease,* Marquart, L., Slavin, J.L., and Fulcher, R.G., Eds., Eagan Press, St. Paul, MN, 2002.
4. Harnack, L., Waltersm S., and Jacobs, J.R., Dietary intake and food sources of whole grains among US children and adolescents: data from the 1994–1996 continuing survey of food intakes by individuals. *J. Am. Diet. Assoc.* 10, 1015, 2003.
5. Lang, R. and Jebb, S.A. Who consumes whole grains, and how much? *Proc. Nutr. Soc.* 62, 123, 2003.
6. Miller, G., Prakash, A. and Decker, E., Whole-grain micronutrients. In *Whole Grain Foods in Health and Disease,* Marquart, L., Slavin, J.L., and Fulcher, R.G., Eds., Eagan Press, St. Paul, MN, 2002.
7. Slavin, J.L., Jacobs, D., and Marquart, L., Grain processing and nutrition. *Crit. Rev. Biotechnology* 21, 49, 2001.
8. Fulcher, R.G. and Rooney-Duke, T.K., Whole-grain structure and organization: implications for nutritionists and processors. In *Whole-Grain Foods in Health and Disease,* Marquart, L., Slavin, J.L., and Fulcher, R.G., Eds., Eagan Press, St. Paul, MN, 2002, 9–45.
9. Liukkonen, K., et al., Processed-induced changes on bioactive compounds in whole grain rye. *Proc. Nutr. Soc.* 62, 117, 2003.
10. Chen, Y., et al., Alkylresorcinols as markers of whole grain wheat and rye in cereal products. *J. Agric. Food Chem.* 52, 8242, 2004.
11. Johnsen, N.F., et al., Intake of whole grains and vegetables determines the plasma enterolactone concentration of Danish Women. *J. Nutr.* 134, 2691, 2004.
12. Anderson, J.W., Whole-grains intake and risk for coronary heart disease. In *Whole Grain Foods in Health and Disease,* Marquart, L., Slavin, J.L., and Fulcher, R.G., Eds., Eagan Press, St. Paul, MN, 2002.
13. Morris, J., Marr, J., and Clayton, D., Diet and heart: a postscript. *Br. Med. J.* 2, 1307, 1977.
14. Brown, L., et al., Cholesterol-lowering effects of dietary fiber: a meta-analysis. *Am. J. Clin. Nutr.* 69, 30, 1999.
15. Jacobs, D.R., et al., Whole-grain intake may reduce the risk of ischemic heart disease death in postmenopausal women: The Iowa Women's Health Study. *Am. J. Clin.Nutr.* 68, 248, 1988.
16. Pietinen, P., et al., Intake of dietary fiber and risk of coronary heart disease in a cohort of Finnish men. The Alpha-Tocopherol, Beta-Carotene Cancer Prevention Study. *Circulation.* 94, 2720, 1996.

17. Rimm, E.B., et al., Vegetable, fruit and cereal fiber intake and risk of coronary heart disease among men. *J. Am. Med. Assoc.* 275, 447, 1996.

18. Liu, S.M., et al., Whole-grain consumption and risk of coronary heart disease: results from the Nurse's Health Study. *Am. J. Clin. Nutr.* 70, 412, 1999.

19. Truswell, A.S., Cereal grains and coronary heart disease. *Eur. J. Clin, Nutr.* 56, 1, 2002.

20. Katz, D.L., et al., Effects of oat and wheat cereals on endothelial responses. *Preventive Medicine.* 33, 476, 2001.

21. Jang, Y., et al., Consumption of whole grain and legume powder reduces insulin demand, lipid peroxidation, and plasma homocysteine concentrations in patients with coronary artery disease: randomized controlled clinical trial. *Arterioscler. Thromb. Vasc. Biol.* 21, 2065, 2001.

22. Pins, J.J., et al., Do whole-grain oat cereals reduce the need for antihypertensive medications and improve blood pressure control? *J. Family Practice.* 51, 353, 2002.

23. Van Dam, R.M., et al., Patterns of food consumption and risk factors for cardiovascular disease in the general Dutch population. *Am. J. Clin. Nutr.* 77, 1156, 2003.

24. Lopez-Garcia, E., et al., Major dietary patterns are related to plasma concentrations of markers of inflammation and endothelial dysfunction. *Am. J. Clin. Nutr.* 80, 1029, 2004.

25. Wu, T., et al., Fructose, glycemic load, and quantity and quality of carbohydrate in relation to plasma C-peptide concentrations in US women. *Am. J. Clin. Nutr.* 80, 1043, 2004.

26. Jensen, M.K., et al., Intakes of whole grains, bran, and germ and the risk of coronary heart disease in men. *Am. J. Clin. Nutr.* 80, 1492, 2004.

27. Van Dam, R.M., et al., Dietary patterns and risk for type 2 diabetes mellitus in US men. *Ann. Int. Med.* 136, 201, 2002.

28. Murtaugh, M.A., et al., Epidemiological support for the protection of whole grains against diabetes. *Proceed. Nutr. Soc.* 62, 143, 2003.

29. Franz, M.J., et al., Evidence-based nutrition principles and recommendations for the treatment and prevention of diabetes and related complications. *Diabetes Care.* 25, 148, 2000.

30. Granfeldt, Y., Hagander, B., and Bjorck, I., Metabolic responses to starch in oat and wheat products. On the importance of food structure, incomplete gelatinization or presence of viscous dietary fibre. *Eur. J. Clin. Nutr.* 49, 189, 1995.

31. Jarvi, A., et al., The influence of food structure on postprandial metabolism in patients with non-insulin-dependent diabetes mellitus. *Am. J. Clin. Nutr.* 61, 837, 1995.

32. Jenkins, D.J.A., et al., Low glycemic response to traditionally processed wheat and rye products: bulgur and pumpernickel bread. *Am. J. Clin. Nutr.* 43, 516, 1986.

33. Jenkins, D.J.A., et al., Whole meal versus whole grain breads: proportion of whole or cracked grain and the glycemic response. *Br. Med. J.* 29, 958, 1988.

34. Heaton, K.W., et al., Particle size of wheat, maize, and oat test meals: effects on plasma glucose and insulin responses and on the rate of starch digestion in vitro. *Am. J. Clin. Nutr.* 47, 675, 1988.

35. Pereira, M.A., et al., Effect of whole grains on insulin sensitivity in overweight hyperinsulinemic adults. *Am. J. Clin. Nutr.* 75, 848, 2002.

36. Juntunen, K.S., et al., Postprandial glucose, insulin, and incretin responses to grain products in healthy subjects. *Am. J. Clin. Nutr.* 75, 254, 2002.

37. Salmeron, J., et al., Dietary fiber, glycemic load, and risk of NIDDM in men. *Diabetes Care.* 20, 545, 1997.

38. Salmeron, J., et al., Dietary fiber, glycemic load, and risk of non-insulin-dependent diabetes mellitus in women. *J. Am. Med. Assoc.* 277, 472, 1997.

39. Meyer, K.A., et al., Carbohydrates, dietary fiber, and incident type 2 diabetes in older women. *Am. J. Clin. Nutr.* 71, 921, 2000.

40. Liu, S., et al., A prospective study of whole grain intake and risk of type 2 diabetes mellitus in U.S. women. *Am. J. Public Health.* 90, 1409, 2000.

41. Fung, T.T., et al., Whole-grain intake and the risk of type 2 diabetes: a prospective study in men. *Am. J. Clin. Nutr.* 76, 535, 2002.

42. Montonen, J., et al., Whole-grain and fiber intake and the incidence of type 2 diabetes. *Am. J. Clin, Nutr.* 77, 622, 2003.

43. Liu, S., Whole-grain foods, dietary fiber, and type 2 diabetes: searching for a kernel of truth. *Am. J. Clin. Nutr.* 77, 527, 2003.

44. McKeown, N.M., et al., Whole-grain intake is favorably associated with metabolic risk factors for type 2 diabetes and cardiovascular disease in the Framingham Offspring Study. *Am. J. Clin. Nutr.* 76, 390, 2002.

45. Juntunen, K.S., et al., High-fiber rye bread and insulin secretion and sensitivity in healthy postmenopausal women. *Am. J. Clin, Nutr.* 77, 385, 2003.

46. Jacobs, D.R., et al., Whole-grain intake and cancer: an expanded review and meta-analysis. *Nutr. Cancer* 30, 85, 1998.

47. Chatenoud, L., et al., Whole-grain food intake and cancer risk. *Int. J. Cancer* 77, 24, 1988.

48. Terry, P., et al., Inverse association between intake of cereal fiber and risk of gastric cardia cancer. *Gastroenterology.* 120, 387, 2001.

49. Kasum, C.M., et al., Dietary risk factors for upper aerodigestive tract cancers. *Int. J. Cancer.* 99:, 267, 2002.

50. Kasum, C.M., et al., Whole grain intake and incident endometrial cancer: the Iowa Women's Health Study. *Nutr. Cancer.* 39, 180, 2001.

51. Schultz, M., et al., Dietary determinants of epithelial ovarian cancer: a review of the epidemiologic literature. *Nutr. Cancer.* 2004;50, 120, 2004.

52. Borriello, S.P., et al., Production and metabolism of lignans by the human faecal flora. *J. App. Bact.* 58, 37, 1985.

53. Adlercreutz, H., et al., Urinary estrogen profile determination in young Finnish vegetarian and omnivorous women, *J. Steroid. Bioch.* 24, 289, 1986.

54. Kilkkinen A., et al., Intake of lignans is associated with serum enterolactone concentration in Finnish men and women. *J. Nutr.* 133, 1830, 2003.

55. Kilkkinen, A., et al., Determinants of serum enterolactone concentration. *Am. J. Clin. Nutr.* 73, 1094, 2001.

56. Jacobs, D.R., et al., Whole grain food intake elevates serum enterolactone. *Br. J. Nutr.* 88, 111, 2002.

57. Vanharanta, M., et al., Risk of cardiovascular disease-related and all-cause death according to serum concentrations of enterolactone. Kuopio Ischaemic Heart Disease Risk Factor Study. *Arch. Int. Med.* 163, 1099, 2003.

58. McIntosh G.H., et al., Whole-grain rye and wheat foods and markers of bowel health in overweight middle-aged men. *Am. J. Clin. Nutr.* 77, 967, 2003.

59. Pereira, M.A., et al., The association of whole grain intake and fasting insulin in a biracial cohort of young adults: the CARDIA study. *CVD Prevention.* 1998 1, 231, 1998.

60. Ludwig, D.S., et al., Dietary fiber, weight gain, and cardiovascular disease risk factors in young adults. *J. Am. Med. Assoc.* 282, 1539, 1999.

61. Newby, P.K., et al., Dietary patterns and changes in body mass index and waist circumference in adults. *Am. J. Clin. Nutr.* 77, 1417, 2003.
62. Koh-Banerjee, P., et al., Changes in whole-grain, bran, and cereal fiber consumption in relation to 8-y weight gain among men. *Am. J. Clin. Nutr.* 80, 1237, 2004.
63. Jacobs, D.R., et al., Is whole-grain intake associated with reduced total and cause-specific death rates in older women? The Iowa Women's Health Study. *Am. J. Publ. Health* 89, 322, 1999.
64. Jacobs, D.R., et al., Fiber from whole grains, but not refined grains, is inversely associated with all cause mortality in older women: the Iowa Women's Health Study. *J. Am. Coll. Nutr.* 19, 326S, 2000.
65. Jacobs, D.R., Meyer, H.E., and Solvoll, K. Reduced mortality among whole grain bread eaters in men and women in the Norwegian County Study. *Eur. J. Clin. Nutr.* 55, 137, 2000.
66. Liu, S., et al., Is intake of breakfast cereals related to total and cause specific mortality in men? *Am. J. Clin. Nutr.* 77, 594, 2003.

CHAPTER 12

Molecular Activities of Vitamin E

Jean-Marc Zingg and Angelo Azzi

CONTENTS

INTRODUCTION

The phytochemicals present in the human diet are unique substances produced during growth and development of plants. In addition to their nutritional role, phytochemicals can have a therapeutic role with health-protective benefits by acting as modifiers of many physiological functions. Whereas most of these dietary phytonutrients are not essential for the human body (flavonoids, polyphenols, most carotenoids, etc.) and are either poorly taken up or their plasma concentration is limited by efficient metabolism and elimination, a few of these molecules are essential for the human body, such as pro-vitamin A (β-carotene), vitamin K (phylloquinone), or vitamin E (tocopherols). The plasma and tissue concentrations of these essential compounds are regulated either via their uptake, metabolism, or retention; concentrations below the physiological normal lead to deficiency syndromes, and concentrations much above that can lead to accumulation and toxicity. Many phytochemicals have the ability to chemically scavenge free radicals and thus act in the test tube as antioxidants, but their main biological activity is by acting as hormones, ligands for transcription factors, modulators of enzymatic activities, or as structural components. In fact, oxidation of these molecules may impair their biological activity, and cellular defense systems exist, which protect these molecules from oxidation.

Vitamin E, which is essential for higher organisms, is present in plants in eight different forms with more or less equal antioxidant potential (α-, β-, γ-, δ-tocopherol/tocotrienols). In higher organisms only α-tocopherol is preferentially retained, suggesting a specific mechanism for the uptake of this analog. Absence of selective α-tocopherol retention leads to vitamin E deficiency with consequent development of disease. In the last 20 years, the route of tocopherol from the diet into the body has been clarified and the proteins involved in the uptake and selective retention of α-tocopherol discovered. Precise cellular functions of α-tocopherol that are independent of its antioxidant/radical scavenging ability have been characterized in recent years. These effects are unrelated to the antioxidant activity of vitamin E and possibly reflect specific interactions of the tocopherols with enzymes, structural proteins, lipids, and transcription factors. Several tocopherol-binding proteins have been cloned, which may mediate the non-antioxidant signaling and cellular functions of vitamin E and its correct intracellular distribution. In the present review, it is suggested that the non-antioxidant activities of the tocopherols represent the main biological reason for the selective retention of α-tocopherol in the body, or vice versa, for the metabolic conversion and consequent elimination of the other tocopherol analogs.

NATURAL AND SYNTHETIC VITAMIN E ANALOGS

Vitamin E was first described by Evans and Bishop as an essential nutrient for reproduction in rats.[1] Vitamin E acts chemically as a free radical chain breaking molecule within the lipid phase (lipoprotein and membranes) and thus protects the organism against the attack of those radicals.[2–4] In the last 20 years, alternative non-antioxidant roles of vitamin E have been proposed, such as the modulation of cellular signaling, enzymatic activity, and gene expression.[5–8] In the studies discussed here, we focus on the different chemical and biological characteristics of the natural and synthetic vitamin E analogs.

Natural Vitamin E

Natural vitamin E comprises 8 different forms, α-, β-, γ-, and δ-tocopherol and α-, β-, γ-, and δ-tocotrienol. The tocotrienols have an unsaturated side chain, whereas the tocopherols contain a phytyl tail with three chiral centers which naturally occur in the RRR configuration (Figure 12.1). The tocopherols are exclusively synthesized in photosynthetic organisms, including higher plants; significant amounts are found in all green tissues and also in seeds.

Tocopherols in plants are generated from the condensation of phytyldiphosphate and homogentisic acid (HGA), followed by cyclization and methylation reactions. Homogentisate phytyltransferase (HPT) performs the first committed step in this pathway, the phytylation of HGA. Tocopherol methylase converts δ- and γ-tocopherol into β- and α-tocopherol, respectively, but β-tocopherol is not accepted as a substrate and, thus, not converted into α-tocopherol.[9] As a consequence, the relative concentration of the different tocopherol analogs in plants depends on the activity of the tocopherol methyltransferases.[10] This was demonstrated by overexpression of γ-tocopherol methyltransferase in *Arabidopsis thaliana* or soybean seeds, shifting oil compositions toward α-tocopherol and thus improving the vitamin E content.[11,12] *Arabidopsis thaliana* plants that lacked all tocopherols due to a deficient tocopherol cyclase activity showed slightly reduced chlorophyll content and photosynthetic quantum yield during photo-oxidative stress, indicating a potential role for tocopherol in maintaining an optimal photosynthesis rate under high-light stress.[13] In

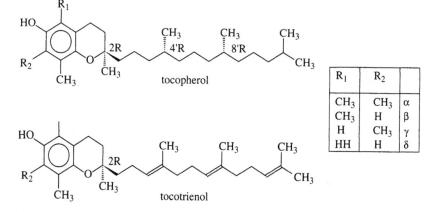

Figure 12.1   Structure of the eight natural tocopherols. In plant tissues, four tocopherols and four tocotrienols are synthesized, all with a side-chain in the natural RRR configuration (here referred to as α-, β-, γ-, δ-tocopherol/tocotrienol). The relative concentration of the tocopherols and tocotrienols depends on the plant species and on the plant tissue.

addition to that, the impaired photoassimilate export and deficient plasmodesmata formation in zea mays deficient in vitamin E (SXD1, Sucrose Export Deficient 1) both suggest that, beyond presumed antioxidant activities, tocopherols or tocopherol breakdown products also function as signal transduction molecules.[14–20]

α-Tocopheryl Phosphate (TP)

Recently, using a novel isolation method, one of the synthetic analogs of tocopherol, the α-tocopheryl phosphate, was shown to occur naturally in foods and in low amounts in animal and human tissues.[21] Sometimes the amounts of α-tocopheryl phosphate are 10- to 30-fold higher than free α-tocopherol (e.g., in chocolate and certain cheeses).[21] In some animal tissues (including humans) the amounts of α-tocopheryl phosphate is of the same order of magnitude as that of α-tocopherol, and in some cases it can be decisively higher (e.g., rat and pig liver).[21] Furthermore, supplementation of the diet of rats with α-tocopheryl phosphate resulted in an increased deposition of α-tocopheryl phosphate and α-tocopherol in liver and adipose tissue.[21] These findings prompt a number of questions, ranging from the possibility that α-tocopheryl phosphate is a reserve form of α-tocopherol to the hypothesis that it may represent an active compound capable of exerting regulatory effects at a cellular level.

Several functions and activities have been suggested for tocopheryl phosphate: induction of hippocampal long term potentiation,[22] protection of mouse skin against ultraviolet-induced damage,[23] activation of cAMP phosphodiesterase,[24] and activation of rat liver phenylalanine hydroxylase.[25] In human THP-1 monocytic leukemia cells and RASMC rat aortic smooth muscle cells, α-tocopheryl phosphate was more potent than α-tocopherol in inhibiting CD36 mRNA and protein expression and cell proliferation.[26] Contrary to α-tocopherol, α-tocopheryl phosphate was cytotoxic to THP-1 cells at high concentrations. The higher potency of α-tocopheryl phosphate may be due to a better uptake of the molecule and to its intracellular hydrolysis, providing more α-tocopherol to sensitive sites. Alternatively, a direct effect of the α-tocopheryl phosphate ester on specific cellular targets may be considered.

It seems possible that α-tocopheryl phosphate acts as a signaling molecule that mediates some of the effects seen on gene expression and cellular signaling. In particular, it seems possible that α-tocopheryl phosphate plays a similar "second messenger" role as known for the phosphorylated forms of phosphatidylinositol, by acting as membrane anchor and by attracting enzymes such as kinases, phosphatases, or NADPH-oxidase to the plasma membrane, leading to their activation/inactivation (Figure 12.2). In fact, low amounts of tocopherol may become phosphorylated and dephosphorylated, suggesting that the inter-conversion may serve some cellular signaling functions. Preliminary results suggest that α-tocopherol can be phosphorylated in HMC-1 mastocytoma cells and primary aortic smooth muscle cells,[27] whereas other studies suggest that α-tocopheryl phosphate can be dephosphorylated in human keratinocytes.[23] Since PKB translocation to the plasma membrane is inhibited by α-tocopherol in HMC-1 mast cells,[28] it appears possible that α-tocopheryl phosphate is mediating these effects.

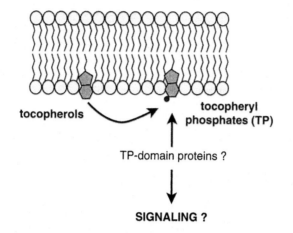

Figure 12.2 Hypothetical cellular signaling by tocopheryl phosphate (TP). The tocopherols become phosphorylated in the membrane by a kinase. Proteins containing a tocopheryl phosphate binding domain (TP-domain) may translocate to TP in the membrane and become activated or inactivated. A phosphatase would inactivate the cellular signaling by removing the phosphate group.

Free Radical Scavenging

The overall antioxidant activity of the natural tocopherols is more or less similar; however, clear individual chemical, physical, and biological effects can be distinguished at the molecular level. The free radical scavenging reactivity has been measured as being in the order of $\alpha > \gamma > \beta > \delta$.[29] The chemical reactivity of the four tocopherols with singlet molecular oxygen ($^1O_2$) has been found to be very low, with $\alpha > \gamma > \delta > \beta$. The physical quenching ability of $^1O_2$ has been measured as being in the order of $\alpha \geq \beta > \gamma > \delta$.[30] The rather complex physical and chemical properties of tocopherols have been extensively reviewed.[31] The *in vivo* biological potency can be summarized with the order of $\alpha \gg \gamma > \delta > \beta$, which is most likely due to the selective retention of α-tocopherol by the liver. α-Tocopheryl phosphate has no antioxidant activity per se, since it is phosphorylated at the chromanol –OH group, which in α-tocopherol is essential for the scavenging of free radicals. However, at extremely high concentrations, it may reduce in *in vitro* systems the propagation of free radicals in membranes from one polyunsaturated fatty acid to another, or possibly interfere with their enzymatic generation.[32]

Commercial Vitamin E

Commercially available vitamin E consists of either a mixture of naturally occurring tocopherols and tocotrienols (from natural sources), RRR-α-tocopherol (formerly called d-α-tocopherol), synthetic α-tocopherol, consisting of the eight possible side-chain stereoisomers at equal amounts (*all rac*-α-tocopherol, formerly called dl-α-tocopherol), or their esters (α-tocopheryl succinate, α-tocopheryl acetate, α-tocopheryl phosphate). α-Tocopheryl succinate and α-tocopheryl acetate are efficiently converted into α-tocopherol by esterases. RRR-α-tocopherol is the most abundant form in plasma (consequently, most supplements contain mainly α-tocopherol), whereas the plasma γ-tocopherol level is only about 10% of that of α-tocopherol despite that a higher amount of γ-tocopherol is often present in the diet. This specificity is the consequence of a selective retention of RRR-α-tocopherol in the body, or vice versa, to the metabolic degradation of the other tocopherols and their elimination. Thus, bioavailability and bioequivalence of the natural forms of vitamin E differ; this is taken into account for the determination of total vitamin E activity in food.

Synthetic Vitamin E Analogs

The synthetic racemic tocopherol mixture contains eight different side-chain isomers, the RRR form (natural) and all the others containing S isomers. Some of these natural and the non-natural tocopherol isomers are excluded from the plasma and secreted with the bile.[33,34] Vitamin E acetate, the most often used analog in food supplements and cosmetic products, is more stable due to its esterification and consequent protection from oxidation. Once in the gut, the esters of vitamin E are split to their unesterified forms under the action of pancreatic and intestinal esterases and only the non-esterified tocopherols are efficiently taken up.[35–38] Other vitamin E analogs, such as α-tocopheryl succinate, the non-hydrolysable RRR-α-tocopherol ether acetic acid analog [2,5,7,8-tetramethyl-2R- (4R,8R,12-trimethyl-tridecyl)chroman-6-yloxyacetic acid (α-TEA)], α-tocopheryl polyethylene glycol 1000 succinate, tocopheryl nicotinate, tocopheryl ferulate, α-tocopheryl oxybutyrate, tretinoin tocopheril, α-tocopheryl phosphate, and trolox, have also been synthesized and their cellular effects investigated.[38–45] These molecules often act as completely novel compounds, are transported differently, and have their own effects on cellular signaling and apoptosis.[42,43,46–50] Since most of these analogs are modified at the tocopherol-6-O position, they do not have any antioxidant activity before their hydrolysis. It is often unknown to what degree these vitamin E analogs enhance or disrupt pathways that are usually used by the natural tocopherols. Vitamin E has also been modified by coupling it with a lipophilic triphenylphosphonium cation through an alkyl linker so that it is targeted to mitochondria and cancer cells.[51]

Tocopherol Metabolites

The selectivity of higher organisms for α-tocopherol has been impressively demonstrated in recent years by analysing the metabolism of vitamin E. Excess α-tocopherol and the other tocopherol analogs are extensively metabolized before excretion. This finding suggests that the organism maintains the correct vitamin E level by selective retention of α-tocopherol, and by specific metabolism of all the other tocopherols and of the excess α-tocopherol. Keep in mind that the tocopherol metabolites can also act as bioactive compounds, which can bind to transcription factors, membrane channels, and enzymes, and modulate their activity (see below).

1. *Simon metabolites:* Initially, two major metabolites of α-tocopherol, the "Simon metabolites" (tocopheronic acid and tocopheronolactone) were described.[52,53] Since they have a shortened side chain and an opened chroman structure, they are often quoted to demonstrate the antioxidant function of α-tocopherol *in vivo*. These metabolites are excreted in the urine as glucuronides or sulfates. The level of these metabolites increases markedly in the urine of healthy volunteers after a daily intake of 2 to 3 g all *rac*-α-tocopherol.

2. *Carboxyethyl hydroxychromans (CEHCs):* Further analysis of the vitamin E metabolism in humans led to the discovery of a novel pathway of tocopherol metabolism.[54] Instead of Simon metabolites, a compound with a shortened side chain but an intact chroman structure, α-carboxyethyl hydroxychroman (α-CEHC), was identified after supplementation with RRR-α-tocopherol.[55] This metabolite is analogous to that of δ-tocopherol found previously in rats[56] and that of γ-tocopherol identified in human urine and proposed as a natriuretic factor.[57] The intact chroman structure of these CEHCs suggests that they are derived from tocopherols which have not reacted as antioxidants. The proposed pathway of side-chain degradation of tocopherol proceeds first via ω- and then β-oxidation.[56] The initial step, the ω-hydroxylation of the side chain is catalyzed by the action of cytochrome P_{450} (CYP)-dependent hydroxylases. Inhibitors of the CYP3A family, like sesamin and ketoconazole, inhibit the formation of γ-CEHC, and dietary intervention with sesame oils in humans leads to increased serum γ-tocopherol levels.[58,59] Induction of CYP3A by rifampicin resulted in an increase of the α-tocopherol metabolites in HepG2 cells.[60] α-CEHC excretion was increased with increasing vitamin E intake after a threshold of plasma α-tocopherol had been exceeded.[55]

CEHC accumulation may mediate anti-inflammatory and antioxidative effects or have other regulatory properties.[61-63] Both α-CEHC and γ-CEHC inhibited microglial PGE(2) production, nitrite production, and reduced iNOS mRNA and protein expression, but neither α- nor γ-tocopherol was effective at inhibiting these cytokine-stimulated inflammatory processes. The metabolite of γ-tocopherol, γ-CEHC has natriuretic activity by inhibition of the 70 pS potassium channel of the thick ascending limb of the loop of Henle, and not by inhibiting the Na^+/K^+-ATPase. The analogous α-tocopherol metabolite showed no inhibition.[64] γ-CEHC inhibits also cyclooxygenase-2 (COX-2) and prostaglandin E2 (PGE2) synthesis in activated macrophages and epithelial cells, events that could change the cellular behavior and affect gene expression.[65,66] In carrageenan-induced inflammation in male Wistar rats, administration of γ-tocopherol or γ-CEHC, but not α-tocopherol, reduces PGE2 synthesis at the site of inflammation, and inhibits leukotriene B4 formation, a potent chemotactic agent synthesized by the 5-lipoxygenase of neutrophils.[67] γ-Tocopherol and γ-CEHC exerted an inhibitory effect on cyclin D1 expression with parallel retardation of cell proliferation.[68] Interestingly, the inhibition of cyclin D1 expression by γ-CEHC was competed for by α-CEHC, suggesting a non-antioxidant mechanism. Another metabolite of vitamin E, 2,2,5,7,8-pentamethyl-6-chromanol (PMCol), has been found to inhibit growth of androgen-sensitive prostate carcinoma cells, which is due to the potent anti-androgenic activity of this compound.[69]

Tocopheryl Quinones

The tocopherols are converted by oxidation to tocopheryl quinones, which upon reduction become tocopheryl hydroquinones. The reduction of α-tocopheryl quinone to α-tocopheryl hydroquinone occurs either via NADPH-cytochrome P_{450} reductase,[70] NAD(P)H:quinone oxidoreductase 1,[71,72] or ascorbate[71]; for the other tocopherols these pathways have not been tested. The tocopheryl hydroquinones can regenerate the tocopheroxyl radical and thus preserve α-tocopherol with different efficiencies ($\alpha > \beta > \gamma$-tocopheryl hydroquinone).[73]

δ-tocopheryl quinone and γ-tocopheryl quinone, but not α-tocopheryl quinone, are cytotoxic to cultured aortic smooth muscle cells, acute lymphoblastic leukemia (ALL) cells and AS52 Chinese hamster ovary cells.[74,75] In HL-60 cells and colon adenocarcinoma WiDr cells, γ-tocopheryl quinone induces apoptosis via caspase-9 activation and cytochrome c release.[76] Furthermore, γ-tocopheryl quinone is highly mutagenic in AS52 cells, whereas α-tocopheryl quinone is not, possibly given an evolutionary advantage to organisms limiting γ-tocopherol, the precursor of γ-tocopheryl quinone.[75] In vivo, dietary α-tocopherol decreased genetic instability in the mouse mutatect tumor model, whereas γ-tocopherol had no effects.[77]

Recently, the tocopherol-associated protein (TAP/SPF) has been described to bind α-tocopheryl quinone, the structure of the complex has been resolved, and it remains to be shown whether any of the effects of the tocopheryl quinones are mediated via the TAP proteins.[78–80]

Occurrence of Vitamin E in Food

The eight vitamin E analogs are widely distributed in nature, and the richest sources are latex lipids (~80 mg/g of latex), followed by edible plant oils (Figure 12.3). Sunflower seeds contain almost exclusively α-tocopherol (59.5 mg/g of oil), oil from soybeans contains the γ-, δ-, and α-tocopherol (62.4, 20.4, and 11.0 mg/g oil), while palm oil contains high concentrations of tocotrienols (17.2 mg/g oil) and α-tocopherol (18.3 mg/g oil).[81] However, the uptake of tocotrienols from the human diet is generally low. As a result of different amounts of the various tocopherols in the oils and the dietary oil preferences in different countries, the plasma and tissue levels of certain tocopherols can be different. In the U.S. the intake of γ-tocopherol is higher than that of α-tocopherol because of the high intake of corn oil. In Europe the intake of α-tocopherol is higher than that of γ-tocopherol because of the high intake of sunflower and olive oils, whereas the higher intake of soybeans in Asia leads to a higher intake of δ-tocopherol. As a consequence, the interpretation of epidemiological studies from different countries may require taking into account the different situations at the start of the supplementation.

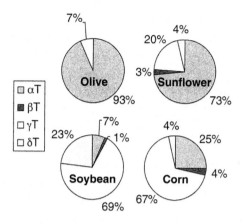

Figure 12.3 Tocopherols distribution in commonly used dietary oils. More than 50% of total vitamin E uptake is derived from dietary oils. Olive and sunflower oils (common to European diet) contain mainly α-tocopherol, corn oils (common to U.S. diet) contain mainly γ-tocopherol, and soy oils (common to Asian diet) contain mainly γ- and δ-tocopherol.[48,81,186,187] With the exception of palm oil, which has high levels of tocotrienols (13% α-tocopherol, 75% tocotrienols), most oils contain only low amounts of tocotrienols (not shown). However, the uptake and retention of tocotrienols from the human diet are generally low.

Plasma and Tissue Concentrations of Vitamin E Analogs

In human plasma, average α-tocopherol concentrations (22 to 28 μM) are about 10 and 100 times higher than γ-tocopherol (2.5 μM) and δ-tocopherol (0.3 μM) concentrations, respectively[48,63] In tissues, the highest contents of α-tocopherol are found in adipose tissue (150 μg/g tissue) and the adrenal gland (132 μg/g tissue), other organs like kidney, heart, or liver contain between 7 and 40 μg/g tissues, and erythrocytes have a relatively low content (2 μg/g tissue).[82,83] These differences in the amount of α-tocopherol suggest tissue-specific mechanisms for enrichment or storage of tissue α-tocopherol. When compared to the other tocopherol analogs, α-tocopherol is found at the highest concentration in plasma and tissues, with the exception of skin, where γ-tocopherol is highest (Figure 12.4). Whereas the proteins involved in selective α-tocopherol retention have been well described in the last 20 years (see below), the mechanisms involved in enrichment of γ-tocopherol in skin or muscle tissue remain unclear.

VITAMIN E UPTAKE AND DISTRIBUTION

Vitamin E Transport from the Intestine

In humans, vitamin E is taken up together with dietary lipids and bile in the proximal part of the intestine, with an average efficiency of about 30%. All four tocopherols are taken up equally, suggesting that, at this level, there is no selectivity. Consequently, a diet rich in γ-tocopherol or δ-tocopherol increases the level

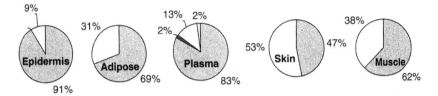

Figure 12.4   Tocopherols distribution in human tissues. With the exception of skin with high levels of γ-tocopherol, α-tocopherol is the most abundant tocopherol in plasma and peripheral tissues. The relative amounts of the tocopherol analogs in human plasma and tissues is determined by their amounts in the diet, their transport by chylomicrons (→ chylomicron tocopherol cycle), the relative affinity of liver α-TTP for the tocopherols (→ α-tocopherol salvage pathway), and the uptake, export and elimination efficiencies in different tissues (see Figure 12.5).

of γ-tocopherol in tissues, albeit in most tissues α-tocopherol is the predominant form. Since competition between the tocopherols occurs, relative tissue levels of tocopherols are dependent on the relative amount of each tocopherol in the diet.[84] The tocopherols are reassembled together with triglycerides, cholesterol, phospholipids, and apolipoproteins into chylomicrons (Figure 12.5). In the course of chylomicron lipolysis, a part of the vitamin E is distributed to peripheral tissues, and the liver with the chylomicron remnants captures the other part (→ chylomicron tocopherol cycle).

α-Tocopherol Salvage Pathway from the Liver

Apparently, the plasma and tissue levels of the four tocopherols reached by chylomicron transport is not sufficient, since α-tocopherol is recycled from the liver by a specific α-tocopherol salvage pathway and mutation of this pathway leads to vitamin E deficiency syndromes (Figure 12.5). The liver α-tocopherol transfer protein (α-TTP) recognizes preferentially α-tocopherol and, thus, plays an important role in determining the plasma vitamin E level (Figure 12.5).[85] Vitamin E is then transported in the blood by VLDL and delivered to peripheral tissues together with triglycerides and cholesteryl esters. Lipolysis of the triglycerides in VLDL by lipoprotein lipase converts VLDL into LDL. The plasma phospholipid transfer protein facilitates the exchange of α-tocopherol between different lipoproteins, such as LDL and HDL, which contain more than 90% of the total serum vitamin E. A part of α-tocopherol is taken up by endothelial cells together with free fatty acids and monoglycerides. Another pathway of α-tocopherol uptake occurs by endocytosis of LDL via the LDL receptor.[86]

In the lungs, HDL is the primary source of vitamin E for type II pneumocytes, and its uptake is regulated by the expression of scavenger receptor SR-BI.[87] In the brain, HDL-associated α-tocopherol is selectively transferred into cells constituting the blood-brain barrier via SR-BI.[88] Similarly, SR-BI transports HDL-associated α-tocopherol coming from the periphery back into the liver, where it is again specifically recognized by α-TTP, recycled, and secreted in VLDL.[89]

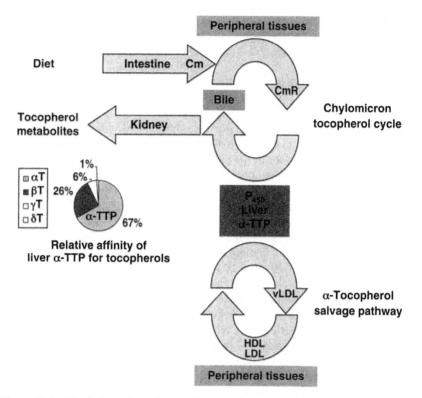

Figure 12.5   Vitamin E uptake and metabolism.

Chylomicron tocopherol cycle and tocopherol metabolism: The four tocopherols are taken up with equal efficiency (average 30%) via the intestine and
distributed to peripheral tissues by chylomicrons (Cm). In the course of chylomicron lipolysis, a part of the vitamin E is distributed to peripheral tissues, and
the liver, with the chylomicron remnants, captures the other part. Chylomicron
remnants (CmR) are taken up by the liver; the α-tocopherol is preferentially
recognized, sorted and secreted with VLDL. The remaining tocopherols and
excess α-tocopherol is metabolized by the cytochrome P450 (CYP3A) enzyme,
a phase I enzyme recognizing foreign compounds.[54] Thus, higher amounts of
β-, γ-, and δ-tocopherols may be recognized as "foreign," possibly because they
have cellular effects that interfere with the normal cellular behavior. Alternatively,
as explained in the text, the metabolites of the tocopherols may have specific
cellular roles. Since they are water soluble, they are cleared by the kidney. Part
of the liver tocopherols (up to 14%) is also secreted with bile, and up to 60%
of biliary α-tocopherol is reabsorbed, thus possibly undergoing a second chylomicron cycle.[188,189]

α-**Tocopherol salvage pathway:** In the liver the α-tocopherol transfer protein
(α-TTP) preferentially recognizes α-tocopherol and incorporates it into VLDL.
During circulation in the blood, VLDL converts to LDL and HDL and delivers
its content including the α-tocopherol to the peripheral tissues. Excess tocopherols are transported back in LDL and HDL to the liver, where they undergo
the next round in this cycle. Since the presence of α-tocopherol in chylomicrons
after dietary uptake is transient, the α-tocopherol salvage pathway may allow
maintaining a continuously increased level of α-tocopherol in plasma and
tissues.

α-TTP is mainly expressed in the liver, and low levels are expressed in the brain,[90] in the retina,[91] lymphocytes, and fibroblasts.[92] Moreover, α-TTP expression in the labyrinthine trophoblast region of the placenta plays an important role in supplying the vitamin to the fetus, which may explain the fetal resorption occurring in rats fed a vitamin E deficient diet.[93] Mutations in α-TTP result in a familial disease, ataxia with vitamin E deficiency (AVED), associated with low levels of α-tocopherol in plasma and symptoms closely resembling those of Friedreich's ataxia.[91,94] By following the plasma concentrations in patients expressing a mutant α-TTP gene, it became clear that the main physiological purpose of the α-tocopherol salvage pathway is to maintain a high and continuous plasma concentration of α-tocopherol.[95,96] Thus, in addition to the dietary availability of α-tocopherol, the expression level of liver α-TTP protein is a critical determinant of the α-tocopherol level in plasma and peripheral cells.

The maintenance of high concentrations of α-tocopherol in plasma protects the lipoproteins (VLDL, LDL, and HDL) from oxidative damage,[97] however, this may not be the only physiological purpose of this pathway. Given the neurodegenerative symptoms caused by a deficient α-TTP gene, it can be assumed that the evolutionary benefit of this pathway is to improve the delivery of α-tocopherol to the brain and the peripheral nervous system by maintaining adequate and continuous plasma levels of α-tocopherol. Other tissues, such as muscle tissues, may be less affected by vitamin E deficiency, since they may receive sufficient tocopherols via the chylomicron tocopherol cycle, whereas the brain may rely more on VLDL/HDL and not chylomicrons as a source for α-tocopherol. This is confirmed in α-TTP knockout mice, in which the delivery of α-tocopherol is lowest for brain and spinal cord (2.5% of normal), whereas other tissues (liver, skin, adrenal gland, muscle, and others) can maintain a significant level of α-tocopherol (10 to 30% of normal), despite the absence of a functional α-tocopherol salvage pathway.[98]

Intracellular Distribution of Vitamin E

Higher concentrations of vitamin E have been described in particular organelles, such as Golgi, lysosomes, and mitochondria. Most α-tocopherol is located in the mitochondrial fractions and in the endoplasmic reticulum, whereas little is found in cytosol and peroxisomes.[46] In mitochondria, more α-tocopherol is found in the inner membrane (83.7%) than in the outer membrane (14.3%).[99] This organelle-specific tocopherol distribution could be explained by specific cellular tocopherol transporters or by selective transport across the endothelial cell layer. In line with this, α- and γ-tocopherols are differently taken up in cultured human endothelial cells, suggesting specific tocopherol transporters and receptors.[100]

In the last decade, several proteins have been described that can bind tocopherols, however, their role in intracellular distribution of the tocopherols remains to be shown in detail.[101–105] A 14.2 kDa tocopherol-binding protein (TBP) was shown to enhance up to ten-fold the transport of α-tocopherol to the mitochondria.[102] Among the tocopherol-binding proteins, only α-TTP and several human tocopherol associated proteins (hTAP1, hTAP2, hTAP3) have been cloned and shown in vitro to bind tocopherol with reasonable affinity (Table 12.1).[78,80,106–109]

Table 12.1 Tocopherol Binding Proteins

Protein	Proposed Function Related to Vitamin E
α-Tocopherol transfer protein (α-TTP)	Tocopherol transport and retention; incorporation of α-tocopherol into VLDL[80,85,105,190]
Tocopherol associated protein 1 (TAP1/SPF)	Tocopherol and phospholipid transport and signaling; regulation of cholesterol synthesis[80,106,107,109,191,192]
Tocopherol associated protein 2 (TAP2)	Tocopherol and phospholipid transport and signaling[109]
Tocopherol associated protein 3 (TAP3)	Tocopherol and phospholipid transport and signaling[109]
Afamin	Tocopherol transport, storage, and neuroprotection[193,194]

The binding of tocotrienols to these proteins has so far only demonstrated for α-TTP, which bound α-tocotrienol with an affinity similar to γ-tocopherol.[78,80,106–109] The hTAP proteins could participate in the intracellular distribution of vitamin E and mediate tocopherol transport to the Golgi apparatus or to the mitochondria.[109,110] Moreover, the role of hTAPs and similar proteins may be that of conferring specificity to the action of the different tocopherols, through recognition and selective transport to enzymes, transcription factors, nuclear receptors such as PXR, or organelles. Indeed, the hTAP1 protein recognizes the different natural tocopherols with different specificity; among the tocopherols, the affinity for γ-tocopherol is highest and it is unknown whether this explains the relatively high concentration of γ-tocopherol in the skin (Figure 12.4).[80] Furthermore, the hTAP proteins bind phospholipids that are involved in cellular signaling, such as phosphatidylglycerol, phosphatidylinositol, phosphatidylserine, and phosphatidic acid, and α-tocopherol can compete with these ligands.[78,109,111] Thus, the hTAP proteins may be involved in the regulation of cellular tocopherol concentration, tocopherol transport, and tocopherol-mediated signaling.[109,112,113]

Cellular Export of Vitamin E

Apart from the α-TTP-mediated VLDL incorporation and export, secretion of α-tocopherol from cultured cells is also mediated by ABCA1 and other not yet characterized processes. ABCA1 is an ATP-binding cassette protein that transports cellular cholesterol and phospholipids to lipid-poor high-density lipoproteins (HDL) and apolipoproteins such as apoA-I.[114] In this study, cells lacking an active ABCA1 pathway markedly increased secretion of α-tocopherol to apoA-I after overexpression of ABCA1.

Possible Biological Reasons for Selective α-Tocopherol Retention

Cells incubated with different natural or synthetic vitamin E analogs show striking differences in cellular responses; since most of the synthetic analogs of vitamin E cannot act as antioxidants, and the natural tocopherols have essentially equal antioxidant activity, such differences can only be explained by specific non-antioxidant interaction of each tocopherol analog with cellular signaling and gene expression pathways.

It is still unexplained why nature selected specifically the α form of tocopherol to selectively increase its plasma and tissue concentration. As discussed below, it can be speculated that α-tocopherol has some specific characteristics (e.g., the fully methylated chromanol-head group may be required for optimal interactions with enzymes or "α-tocopherol receptors"). In addition, the protection of the –OH group of α-tocopherol by the nearby methyl groups may render its oxidation more difficult and, once oxidation had taken place, the resulting radical may be long lived enough to permit reduction and restoration of the original molecule. On the other hand, the β-, γ-, and δ- tocopherols and the α-, β-, γ-, and δ- tocotrienols and their metabolites, may have biological effects at higher concentrations that may interfere with normal cellular processes, so that they need to be specifically recognized, metabolized by the liver, and later eliminated. However, adverse effects of these tocopherols analogs are difficult to assess experimentally, since in the normal setting they are continuously eliminated from the body. It is furthermore unknown whether these tocopherols analogs and their metabolites have an essential cellular function at low concentration; since they are eliminated through the bile, they may in fact exert a direct beneficial effect in the gastrointestinal tract.[115]

A unique feature of α-tocopherol is the location of the reactive –OH group between two methyl groups; after reacting with a lipid peroxide the unpaired electron can delocalize over the fully substituted chromanol ring, which is known to increase its stability and chemical reactivity.[31,116] As a consequence, α-tocopherol and α-tocotrienol, but not the other forms of tocopherol, can reduce *in vitro* Cu(II) to give Cu(I) together with α-tocopheryl quinones and α-tocotrienyl quinone, respectively, and they can exert pro-oxidant effects in the oxidation of methyl linoleate in SDS micelles.[117] However, albeit α-tocotrienol has higher antioxidant activity than α-tocopherol in membranes,[118] only α-tocopherol is retained; this suggests that the liver α-TTP protein evolved to recognize and retain α-tocotrienol inefficiently, since the presence of continuously increased levels of α-tocotrienol in serum possibly would interfere with normal cellular reactions. In addition, since α-tocotrienol is taken up into certain cells in culture with higher efficiency than α-tocopherol, it may exert biological effects already at lower effective concentrations. In line with this model, it was shown that α-tocotrienol was more potent in many cellular reactions than α-tocopherol, such as in reducing cholesterol levels by inhibition of HMG-CoA reductase activity and its mRNA translation, by decreasing the secretion of apoB and increasing its proteasomal degradation,[119–123] or by blocking glutamate-induced death.[124] During ischemia/reperfusion, tocotrienol restored both 20S- and 26S-proteasome activities and significantly inhibited the phosphorylation of cSrc.[125] Among the vitamin E analogs examined, α-tocotrienol exhibited the most potent neuroprotective actions in rat striatal cultures.[126] Furthermore, the efficacy of tocotrienol for reduction of VCAM-1 expression and adhesion of THP-1 cells and monocytes to HUVECs was ten-fold higher than that of α-tocopherol, and this was explained by a higher cellular uptake of α-tocotrienol in these cells.[127–129] The tocotrienols were more potent than the tocopherols in inhibiting proliferation and inducing apoptosis in the estrogen-responsive MCF7 and estrogen-nonresponsive

MDA-MB-435 human breast cancer cell lines in culture,[130] as well as in preneoplastic and neoplastic mouse mammary epithelial cells.[131–133] The α- and γ-tocotrienols were effective against transplantable murine tumors (sarcoma 180, Ehrlich carcinoma, and IMC carcinoma), whereas α-tocopherol had only a slight effect.[134] Recent studies showed that tocotrienol-induced apoptosis results from the activation of specific intracellular cysteine proteases (caspases) associated with death receptor activation and signal transduction.[135]

MOLECULAR ACTION OF VITAMIN E

The molecular effects of the different tocopherols can be classified as either antioxidant, pro-oxidant, antialkylating or non-antioxidant. In the following we will focus on non-antioxidant cellular properties of vitamin E (Table 12.2); the other activities of vitamin E have been extensively reviewed.[6,8,136–142]

Modulation of Enzymatic Activity by Vitamin E

Over the last decade, vitamin E has been shown to have specific effects on cellular signalling and gene regulation, modulating cellular events such as proliferation, apoptosis/survival, migration, and adhesion.[5,6,8,143] The main effects on signalling at the enzymatic level are inhibition of protein kinase C (PKC) activity,[144,145] activation of protein phosphatase 2A (PP2A),[146] inhibition of protein kinase B (PKB) activity,[28,147] modulation of phospholipase A2 activity,[148] and inhibition of cyclooxygenase-2 activity.[149] The crystal structure of phospholipase A2 with the inhibitory vitamin E is a strong example of non-antioxidant vitamin E-enzyme interaction with regulatory function.[150] In many situations, only α-tocopherol has been checked, and it

Table 12.2 Molecular Activities of Vitamin E

Activity of Vitamin E	Target of Vitamin E
Antioxidant activity	Scavenging of reactive oxygen species (ROS) and reactive nitrogen species (RNS)[6,8,136–142]
Prooxidant activity	Lipid peroxidation[197,198]
Anti-alkylating activity	Scavenging of reactive oxygen species (ROS) and reactive nitrogen species (RNS) [199–201]
Non-antioxidant activity	Reviewed in [5,6,8,166]
Modulation of enzymatic activity	Protein kinase C (PKC)[145]
	Protein kinase B (PKB)[28,147]
	Protein tyrosine phosphorylation[202,203]
	Protein phosphatase 2A (PP2A)[50,145]
	Phospholipase A2[148,204]
	Cyclooxygenase[149]
	5-Lipoxygenase[205]
	Glutathione S-transferase[206]
	NADPH-oxidase[207]
Modulation of gene expression	More than 32 genes affected[8,152,208,209]

is unclear whether other tocopherols or tocotrienols work equally well. In some studies, differences between the natural tocopherols were found, for example, δ-tocopherol was more potent than α-tocopherol in inhibiting proliferation and induction of apoptosis in human mast cells as well as human hepatoma cells (HepG2).[28,151] In other experiments, the effects of vitamin E have been only tested in the test tube and need to be confirmed *in vivo*. Since the natural tocopherol analogs have essentially equal antioxidant activity, differences seen in cellular reactions can only be explained by non-antioxidant activities of the tocopherols. In summary, it can be assumed that the tocopherols modulate cellular behavior by specific interactions with enzymes, structural proteins, lipids, and transcription factors.

Modulation of Gene Expression by Vitamin E

Several genes have been described as being modulated by tocopherol.[8,152] However, how the tocopherols can modulate gene expression is not yet clearly resolved and, indeed, may involve several different antioxidant and non-antioxidant molecular mechanisms. To explain all the effects seen at the level of gene expression, several regulatory pathways have to be considered:

1. The tocopherols may influence gene expression by direct modulation of the activity of specific transcription factors in a non-antioxidant fashion, for example, via the pregnane X receptor (PXR),[153] possibly other nuclear receptors such as the peroxisome proliferators activated receptors (PPARs), orphan nuclear receptors, or via one of three human tocopherol associated proteins, hTAPs, recently reported to modulate gene expression.[109,154] As an example, the PPARγ mRNA and protein expression in the SW480 colon cancer cells is upregulated by α- and γ-tocopherol.[155] Furthermore, vitamin E activates CYP3A4 and CYP3A5 mRNA expression via the human pregnane X receptor (PXR) in a tocopherol-specific manner: α-tocopherol activates weakly, whereas β-, γ-, and δ-tocopherol and the tocotrienols lead to stronger induction, whereas the tocopherol metabolic products do not activate.[156] PXR is involved in the drug hydroxylation and elimination pathways, and it activates genes such as cytochromes P_{450} (CYP) (e.g., CYP3A and some ABC transporters).[153] A physiological explanation for the selective retention of α-tocopherol and the elimination of all the others tocopherol analogs could thus be the absence of strong PXR activation by α-tocopherol and the consequent absence of induction of enzymes involved in its metabolism. In the liver, α-tocopherol may be specifically sorted by α-TTP into vesicles destined for incorporation into VLDL. Only when the level of α-tocopherol exceeds the capacity of α-TTP may transport to the metabolic enzymes occur. The other tocopherols are not retained by α-TTP, activate PXR, and then become metabolized and eliminated by CYP3A. Tocopherol-binding proteins, such as α-TTP and hTAPs, could mediate these effects by generating specificity to the metabolism and action of the tocopherols.
2. The tocopherols can change the activity of transcription factors and signal transduction pathways by modulating enzymes, such as protein phosphatase 2A (PP2A), protein kinase C α (PKCα), tyrosine kinases, protein kinase B (PKB), the phospholipase A_2, 5-lipoxygenase, and cyclooxygenase 2, which could indirectly influence gene expression.[8] Phosphorylation of the nuclear receptor, RXRα,

was inhibited by α-tocopherol as a result of PKC inhibition, leading to increased expression of the CRABP-II gene.[157] In the absence of PMA, α-tocopherol led to activation of AP-1, whereas in the presence of PMA inhibition occurred, whereas β-tocopherol had no effect.[158,159]

3. The tocopherols may also influence gene expression by binding to proteins like hTAP, which may act as "molecular chaperones," which generate specificity to the action of the tocopherols. These proteins may regulate tocopherol access to specific enzymes and transcription factors or control the level of "free" tocopherol. The hTAPs modulate *in vitro* the activity of recombinant phosphatidylinositol-3-kinase and α-tocopherol modulates kinase activity in an hTAP-dependent manner, possibly by competition with phosphatidylinositol. Thus, by modulating the intracellular targeting of these ligands to enzymes and organelles, the hTAPs may influence the activity of lipid-dependent enzymes.[109,112]

4. The tocopherols may be metabolized to bioactive compounds, such as Simon metabolites, CEHCs, or tocopheryl quinones (see sections on tocopherol metabolites and tocopheryl quinones), which can bind to transcription factors and enzymes and modulate their activity.

5. Tocopherols may act as a precursor for the synthesis of tocopheryl phosphates, which would be the real effector molecules by influencing cellular events such as signal transduction, apoptosis/survival, and gene expression (see section on α-tocopherol phosphate).

PREVENTIVE EFFECTS OF VITAMIN E

Prevention of Vitamin E Deficiency Syndromes

It should be noted that vitamin E deficiency in humans with full neurological symptoms is rare and usually is the consequence of mutations of the α-tocopherol transfer protein (α-TTP) leading to ataxia with vitamin E deficiency (AVED). These patients with plasma α-tocopherol concentrations below 2.2 μM are affected by ataxia, loss of neurons, retinal atrophy, massive accumulation of lipofuscin in neurons, and retinitis pigmentosa.[48,94,160] These symptoms are similar to those of Friedreich's ataxia, a disease caused by defective expression of frataxin and the consequent increased mitochondrial oxidative damage and cell death, which is also beneficially influenced by vitamin E.[91] In Friedreich's ataxia fibroblasts cell damage can be more efficiently prevented by using a mitochondria-targeted vitamin E derivative (MitoVit E) than by using α-tocopherol.[161] Similar to that, the combined coenzyme Q(10) (400 mg/d) and vitamin E (2100 IU/d) therapy of patients with Friedreich's ataxia over 47 months resulted in sustained improvement in mitochondrial energy synthesis that was associated with a slowing of the progression of certain clinical features and a significant improvement in cardiac function.[162]

What is more frequent than vitamin E deficiency is vitamin E insufficiency, with suboptimal vitamin E supply in the diet or inefficient uptake and distribution of vitamin E. The normal average plasma concentration of vitamin E is 23.2 μM; a plasma level below 11.6 μM is regarded as deficient.[163] Certain diseases, like abetalipoproteinemia, chronic cholestatic liver disease, cystic fibrosis, chronic pancreatitis, progressive systemic sclerosis, short-bowel syndrome, or several other lipid malabsorption syndromes are associated with a low efficiency of vitamin E uptake, leading to similar symptoms as in AVED. These symptoms clearly can be prevented and, in some situations, reversed by supplemental vitamin E.[163]

Prevention of Atherosclerosis, Cancer, Fibrosis, and Neurodegenerative Diseases by Vitamin E

Several epidemiological studies have shown a preventive effect of α-tocopherol on atherosclerosis and other diseases (Table 12.3).[5,6,8,164-168] However, a more recent clinical trial (the HOPE study) has failed to demonstrate the clinical utility for α-tocopherol in the advanced cardiovascular patient.[169] In a recent follow-up study (HOPE-TOO), it was concluded that long-term vitamin E supplementation does not prevent cancer or major cardiovascular events in patients with vascular disease or diabetes mellitus.[170] Thus, vitamin E may act more as a long-term preventive agent, and in patients with advanced disease (as in the HOPE study), its effect may not be evident. In these studies, vitamin E in lipoproteins (mainly LDL) and also in the subendothelial space has been hypothesized to play a central role in reducing athero-genesis by preventing lipid peroxidation and consequent lesion development.[142] However, the non-antioxidant activities of vitamin E suggest alternative molecular pathways that could be involved in the preventive effects of vitamin E.

Other studies have described preventive effects of vitamin E supplementation against certain types of cancer,[171-175] fibrotic disease,[176,177] or neurodegenerative diseases such as Alzheimer's[178,179] or Parkinson's disease[180] (reviewed by J.M. Zingg and

Table 12.3 Preventive Effects of Vitamin E on Several Diseases

Disease	Proposed Mechanisms
Atherosclerosis	Inhibition of smooth muscle cell proliferation; inhibition of inflammatory processes; modulation of gene expression; prevention of lipid and lipoprotein (LDL) oxidation[5,6,8,164,167,168,170]
Cancer (prostate, breast, colorectal, lung)	Inhibition of cell proliferation by modulation of signal transduction and gene expression; induction of apoptosis[171-175]
Fibrotic diseases	Inhibition of oxidative stress; modulation of signal transduction and gene expression[176,177]
Neurodegenerative diseases (Alzheimer's, Parkinson's)	Prevention of cell death; inhibition of oxidative stress; modulation of signal transduction and gene expression[178-180]
Disorders associated with vitamin E deficiency (ataxia with vitamin E deficiency (AVED), chronic cholestatic liver disease, cystic fibrosis, chronic pancreatitis, short bowel syndrome, progressive systemic sclerosis, abetalipoproteinaemia)	Prevention of cell death; inhibition of oxidative stress; modulation of signal transduction and gene expression[163,195,196]

A. Azzi[8] and A. Azzi et al.[166]). All studies are based on the finding that vitamin E levels can be increased by extra dietary supplementation, implying that pathways involved in vitamin E uptake and body distribution are often not saturated. Plasma vitamin E concentrations could vary in different individuals as a consequence of consumption by excessive production of oxidants or of a deficient uptake and transport into plasma and tissues. Polymorphisms of tocopherol binding proteins, like hTTP or hTAPs and others, and their cellular expression levels, could be the reason for the individual vitamin E uptake and response and could explain the differential susceptibility to disorders such as atherosclerosis, certain cancers, and neurodegenerative diseases.[181] Thus, further studies about proteins possibly involved in tocopherol uptake, cellular distribution, and signaling are required to explain the beneficial function of vitamin E in several degenerative diseases, ranging from cardiomyopathy to Alzheimer's, Parkinson's, and several ataxias with unknown etiologies. These diseases have in common that they often show mitochondrial impairment, and reduced levels of mitochondrial vitamin E may accelerate this process.[180,182–185]

CONCLUSIONS

The results summarized in this review strongly indicate that each natural and synthetic vitamin E analog can have its specific biological effects, and these effects are often not linked to its antioxidant activity. Thus, the term vitamin E should only be used when also specifying the exact analog composition, in particular when describing the activities of vitamin E in experimental settings. At the molecular level, the different vitamin E analog are modulating signal transduction and gene expression, most likely by specific interaction of each vitamin E analog with specific proteins and cellular structures, and not by acting primarily as antioxidants. As a consequence it is reasonable to assume that the basis of the selective retention of α-tocopherol is found in some of the non-antioxidant activities of this analog, or that the other tocopherol analog at higher concentrations perform biological activities that interfere with the normal cellular performance. Both of these molecular activities may explain the health-promoting effects described for vitamin E.

ACKNOWLEDGMENTS

This study was made possible thanks to the support of the Swiss National Science Foundation.

REFERENCES

1. Evans, H.M. and Bishop, K.S., On the existence of a hitherto unrecognized dietary factor essential for reproduction, *Science* 56, 650–651, 1922.

2. Ingold, K.U., Bowry, V.W., Stocker, R., and Walling, C., Autoxidation of lipids and antioxidation by alpha-tocopherol and ubiquinol in homogeneous solution and in aqueous dispersions of lipids: unrecognized consequences of lipid particle size as exemplified by oxidation of human low density lipoprotein, *Proc. Natl. Acad. Sci. U.S.A.* 90 (1), 45–49, 1993.

3. Niki, E., Interaction of ascorbate and alpha-tocopherol, *Ann. N. Y. Acad. Sci.* 498, 186-99, 1987.

4. Smith, D., O'Leary, V.J., and Darley-Usmar, V.M., The role of alpha-tocopherol as a peroxyl radical scavenger in human low density lipoprotein, *Biochem. Pharmacol.* 45 (11), 2195–2201, 1993.

5. Ricciarelli, R., Zingg, J.M., and Azzi, A., Vitamin E: protective role of a Janus molecule, *FASEB. J.* 15 (13), 2314–2325, 2001.

6. Brigelius-Flohe, R., Kelly, F.J., Salonen, J.T., Neuzil, J., Zingg, J.M., and Azzi, A., The European perspective on vitamin E: current knowledge and future research, *Am. J. Clin. Nutr.* 76 (4), 703–716, 2002.

7. Rimbach, G., Minihane, A.M., Majewicz, J., Fischer, A., Pallauf, J., Virgli, F., and Weinberg, P.D., Regulation of cell signalling by vitamin E, *Proc. Nutr. Soc* 61 (4), 415–425, 2002.

8. Zingg, J.M. and Azzi, A., Non-antioxidant activities of vitamin E, *Cur. Med. Chem.* 11 (9), 1113–1133, 2004.

9. Koch, M., Lemke, R., Heise, K.P., and Mock, H.P., Characterization of gamma-tocopherol methyltransferases from Capsicum annuum L and Arabidopsis thaliana, *Eur. J. Biochem.* 270 (1), 84–92, 2003.

10. Ajjawi, I. and Shintani, D., Engineered plants with elevated vitamin E: a nutraceutical success story, *Trends Biotechnol.* 22 (3), 104–107, 2004.

11. Shintani, D. and DellaPenna, D., Elevating the vitamin E content of plants through metabolic engineering, *Science* 282 (5396), 2098–2100, 1998.

12. Van Eenennaam, A.L., Lincoln, K., Durrett, T.P., Valentin, H.E., Shewmaker, C.K., Thorne, G.M., Jiang, J., Baszis, S.R., Levering, C.K., Aasen, E.D., Hao, M., Stein, J.C., Norris, S.R., and Last, R.L., Engineering vitamin E content: from Arabidopsis mutant to soy oil, *Plant Cell* 15 (12), 3007–3019, 2003.

13. Porfirova, S., Bergmuller, E., Tropf, S., Lemke, R., and Dormann, P., Isolation of an Arabidopsis mutant lacking vitamin E and identification of a cyclase essential for all tocopherol biosynthesis, *Proc. Natl. Acad. Sci. U.S.A.* 99 (19), 12495–12500, 2002.

14. Demmig-Adams, B. and Adams, W.W., 3rd, Antioxidants in photosynthesis and human nutrition, *Science* 298 (5601), 2149–2153, 2002.

15. Sattler, S.E., Cahoon, E.B., Coughlan, S.J., and DellaPenna, D., Characterization of tocopherol cyclases from higher plants and cyanobacteria. Evolutionary implications for tocopherol synthesis and function, *Plant Physiol.* 132 (4), 2184–2195, 2003.

16. Hofius, D. and Sonnewald, U., Vitamin E biosynthesis: biochemistry meets cell biology, *Trends Plant Sci.* 8 (1), 6–8, 2003.

17. Hofius, D., Hajirezaei, M.R., Geiger, M., Tschiersch, H., Melzer, M., and Sonnewald, U., RNAi-mediated tocopherol deficiency impairs photoassimilate export in transgenic potato plants, *Plant. Physiol.* 135 (3), 1256–1268, 2004.

18. Munne-Bosch, S. and Falk, J., New insights into the function of tocopherols in plants, *Planta* 218 (3), 323–326, 2004.

19. Munne-Bosch, S., Linking tocopherols with cellular signaling in plants, *New Phytol.* 166 (2), 363–366, 2005.

20. Schneider, C., Chemistry and biology of vitamin E, *Mol Nutr Food Res* 49 (1), 7–30, 2005.

21. Ogru, E., Gianello, R., Libinaki, R., Smallridge, A., Bak, R., Geytenbeck, S., Kannar, D., and West, S., in *Free Radicals and Oxidative Stress: Chemistry, Biochemistry and Pathophysiological Implications,* Galaris, D., Ed., Medimond, Bologna, 2003, pp. 109–113.

22. Xie, Z. and Sastry, B.R., Induction of hippocampal long-term potentiation by alpha-tocopherol, *Brain Res.* 604 (1–2), 173–179, 1993.

23. Nakayama, S., Katoh, E.M., Tsuzuki, T., and Kobayashi, S., Protective effect of alpha-tocopherol-6-O-phosphate against ultraviolet B-induced damage in cultured mouse skin, *J. Invest. Dermatol.* 121 (2), 406–411, 2003.

24. Sakai, T., Okano, T., Makino, H., and Tsudzuki, T., Activation of cyclic AMP phosphodiesterase by a new vitamin E derivative, *J. Cyclic Nucleotide Res.* 2 (3), 163–170, 1976.

25. Abita, J.P., Parniak, M., and Kaufman, S., The activation of rat liver phenylalanine hydroxylase by limited proteolysis, lysolecithin, and tocopherol phosphate. Changes in conformation and catalytic properties, *J. Biol. Chem.* 259 (23), 14560–14566, 1984.

26. Munteanu, A., Zingg, J.M., Ogru, E., Libinaki, R., Gianello, R., West, S., Negis, Y., and Azzi, A., Modulation of cell proliferation and gene expression by alpha-tocopheryl phosphates: relevance to atherosclerosis and inflammation, *Biochem. Biophys. Res. Commun.* 318 (1), 311–316, 2004.

27. Negis, Y., Zingg, J.M., Ogru, E., Gianello, R., Libinaki, R., and Azzi, A., On the existence of cellular tocopheryl phosphate, its synthesis, degradation and cellular roles: a hypothesis, *IUBMB Life* 57 (1), 23–25, 2005.

28. Kempna, P., Reiter, E., Arock, M., Azzi, A., and Zingg, J.M., Inhibition of HMC-1 mast cell proliferation by vitamin E: involvement of the protein kinase B pathway, *J. Biol. Chem.* 279 (49), 50700–50709, 2004.

29. Lien, E.J., Ren, S., Bui, H.H., and Wang, R., Quantitative structure-activity relationship analysis of phenolic antioxidants, *Free Radic. Biol. Med.* 26 (3–4), 285–294, 1999.

30. Kaiser, S., Di Mascio, P., Murphy, M.E., and Sies, H., Physical and chemical scavenging of singlet molecular oxygen by tocopherols, *Arch. Biochem. Biophys.* 277 (1), 101–108, 1990.

31. Kamal-Eldin, A. and Appelqvist, L.A., The chemistry and antioxidant properties of tocopherols and tocotrienols, *Lipids* 31 (7), 671–701, 1996.

32. Rezk, B.M., Haenen, G.R., Van Der Vijgh, W.J., and Bast, A., The extraordinary antioxidant activity of vitamin E phosphate, *Biochim. Biophys. Acta* 1683 (1–3), 16–21, 2004.

33. Traber, M.G. and Kayden, H.J., Preferential incorporation of alpha-tocopherol vs gamma-tocopherol in human lipoproteins, *Am. J. Clin. Nutr.* 49 (3), 517–526, 1989.

34. Brigelius-Flohe, R. and Traber, M.G., Vitamin E: function and metabolism, *FASEB. J.* 13 (10), 1145–1155, 1999.

35. Muller, D.P., Manning, J.A., Mathias, P.M., and Harries, J.T., Studies on the intestinal hydrolysis of tocopheryl esters, *Int. J. Vitam. Nutr. Res.* 46 (2), 207–210, 1976.

36. Mathias, P.M., Harries, J.T., Peters, T.J., and Muller, D.P., Studies on the in vivo absorption of micellar solutions of tocopherol and tocopheryl acetate in the rat: demonstration and partial characterization of a mucosal esterase localized to the endoplasmic reticulum of the enterocyte, *J. Lipid. Res.* 22 (5), 829–837, 1981.

37. Nierenberg, D.W., Lester, D.C., and Colacchio, T.A., Determination of tocopherol and tocopherol acetate concentrations in human feces using high-performance liquid chromatography, *J. Chromatogr.* 413, 79–89, 1987.

38. Lauridsen, C., Hedemann, M.S., and Jensen, S.K., Hydrolysis of tocopheryl and retinyl esters by porcine carboxyl ester hydrolase is affected by their carboxylate moiety and bile acids, *J. Nutr. Biochem.* 12 (4), 219–224, 2001.

39. Sokol, R.J., Heubi, J.E., Butler-Simon, N., McClung, H.J., Lilly, J.R., and Silverman, A., Treatment of vitamin E deficiency during chronic childhood cholestasis with oral d-alpha-tocopheryl polyethylene glycol-1000 succinate, *Gastroenterology* 93 (5), 975–985, 1987.

40. Sokol, R.J., Butler-Simon, N., Conner, C., Heubi, J.E., Sinatra, F.R., Suchy, F.J., Heyman, M.B., Perrault, J., Rothbaum, R.J., Levy, J. et al., Multicenter trial of d-alpha-tocopheryl polyethylene glycol 1000 succinate for treatment of vitamin E deficiency in children with chronic cholestasis, *Gastroenterology* 104 (6), 1727–1735, 1993.

41. Mori, Y., Hatamochi, A., Takeda, K., and Ueki, H., Effects of tretinoin tocoferil on gene expression of the extracellular matrix components in human dermal fibroblasts in vitro, *J. Dermatol. Sci.* 8 (3), 233–238, 1994.

42. Funasaka, Y., Chakraborty, A.K., Komoto, M., Ohashi, A., and Ichihashi, M., The depigmenting effect of alpha-tocopheryl ferulate on human melanoma cells, *Br. J. Dermatol.* 141 (1), 20–29, 1999.

43. Birringer, M., EyTina, J.H., Salvatore, B.A., and Neuzil, J., Vitamin E analogues as inducers of apoptosis: structure-function relation, *Br. J. Cancer* 88 (12), 1948–1955, 2003.

44. Kline, K., Yu, W., and Sanders, B.G., Vitamin E and breast cancer, *J. Nutr.* 134 (12 Suppl), 3458S–3462S, 2004.

45. Wu, Y., Zu, K., Ni, J., Yeh, S., Kasi, D., James, N. S., Chemler, S., and Ip, C., Cellular and molecular effects of alpha-tocopheryloxybutyrate: lessons for the design of vitamin E analog for cancer prevention, *Anticancer Res.* 24 (6), 3795–3802, 2004.

46. Drevon, C.A., Absorption, transport and metabolism of vitamin E, *Free. Radic. Res. Commun.* 14 (4), 229–246, 1991.

47. Cheeseman, K.H., Holley, A.E., Kelly, F.J., Wasil, M., Hughes, L., and Burton, G., Biokinetics in humans of RRR-alpha-tocopherol: the free phenol, acetate ester, and succinate ester forms of vitamin E, *Free Radic. Biol. Med.* 19 (5), 591–598, 1995.

48. Burton, G.W., Traber, M.G., Acuff, R.V., Walters, D.N., Kayden, H., Hughes, L., and Ingold, K.U., Human plasma and tissue alpha-tocopherol concentrations in response to supplementation with deuterated natural and synthetic vitamin E, *Am. J. Clin. Nutr.* 67 (4), 669–684, 1998.

49. Neuzil, J., Weber, T., Terman, A., Weber, C., and Brunk, U.T., Vitamin E analogues as inducers of apoptosis: implications for their potential antineoplastic role, *Redox. Rep.* 6 (3), 143–151, 2001.

50. Neuzil, J., Weber, T., Schroder, A., Lu, M., Ostermann, G., Gellert, N., Mayne, G.C., Olejnicka, B., Negre-Salvayre, A., Sticha, M., Coffey, R.J., and Weber, C., Induction of cancer cell apoptosis by alpha-tocopheryl succinate: molecular pathways and structural requirements, *FASEB. J.* 15 (2), 403–415., 2001.

51. Smith, R.A.J., Porteous, C.M., Gane, A.M., and Murphy, M.P., Delivery of bioactive molecules to mitochondria *in vivo*, *Proc. Natl. Acad. Sci. U.S.A.* 100 (9), 5407–5412, 2003.

52. Simon, E.J., The metabolism of vitamin E. II. Purification and characterization of urinary metabolites of alpha-tocopherol., *J. Biol. Chem.* 221, 807–817, 1956.

53. Simon, E.J., Gross, C.S., and Milhorat, A.T., The metabolism of vitamin E. The absorption and excretion of d-α-tocopheryl-5-methyl-C14-succinate, *J. Biol. Chem.* 221, 797–805, 1956.

54. Brigelius-Flohe, R., Vitamin E and drug metabolism, *Biochem. Biophys. Res. Commun.* 305 (3), 737–740, 2003.

55. Schultz, M., Leist, M., Petrzika, M., Gassmann, B., and Brigelius-Flohe, R., Novel urinary metabolite of alpha-tocopherol, 2,5,7,8-tetramethyl-2(2'-carboxyethyl)-6-hydroxychroman, as an indicator of an adequate vitamin E supply?, *Am. J. Clin. Nutr.* 62 (6 Suppl), 1527S–1534S, 1995.

56. Chiku, S., Hamamura, K., and Nakamura, T., Novel urinary metabolite of d-delta-tocopherol in rats, *J. Lipid. Res.* 25 (1), 40–48, 1984.

57. Wechter, W.J., Kantoci, D., Murray, E.D., Jr., D'Amico, D.C., Jung, M.E., and Wang, W.H., A new endogenous natriuretic factor: LLU-alpha, *Proc. Natl. Acad. Sci. U.S.A.* 93 (12), 6002–6007, 1996.

58. Lemcke-Norojarvi, M., Kamal-Eldin, A., Appelqvist, L.A., Dimberg, L.H., Ohrvall, M., and Vessby, B., Corn and sesame oils increase serum gamma-tocopherol concentrations in healthy Swedish women, *J. Nutr.* 131 (4), 1195–1201, 2001.

59. Parker, R.S., Sontag, T.J., and Swanson, J.E., Cytochrome P4503A-dependent metabolism of tocopherols and inhibition by sesamin, *Biochem. Biophys. Res. Commun.* 277 (3), 531–534, 2000.

60. Birringer, M., Drogan, D., and Brigelius-Flohe, R., Tocopherols are metabolized in HepG2 cells by side chain omega-oxidation and consecutive beta-oxidation, *Free Radic. Biol. Med.* 31 (2), 226–232, 2001.

61. Himmelfarb, J., Kane, J., McMonagle, E., Zaltas, E., Bobzin, S., Boddupalli, S., Phinney, S., and Miller, G., Alpha and gamma tocopherol metabolism in healthy subjects and patients with end-stage renal disease, *Kidney Int.* 64 (3), 978–991, 2003.

62. Grammas, P., Hamdheydari, L., Benaksas, E.J., Mou, S., Pye, Q.N., Wechter, W.J., Floyd, R.A., Stewart, C., and Hensley, K., Anti-inflammatory effects of tocopherol metabolites, *Biochem. Biophys. Res. Commun.* 319 (3), 1047–1052, 2004.

63. Hensley, K., Benaksas, E.J., Bolli, R., Comp, P., Grammas, P., Hamdheydari, L., Mou, S., Pye, Q.N., Stoddard, M.F., Wallis, G., Williamson, K.S., West, M., Wechter, W.J., and Floyd, R.A., New perspectives on vitamin E: gamma-tocopherol and carboxyethylhydroxychroman metabolites in biology and medicine, *Free Radic Biol Med* 36 (1), 1–15, 2004.

64. Murray, E.D., Jr., Wechter, W.J., Kantoci, D., Wang, W.H., Pham, T., Quiggle, D.D., Gibson, K.M., Leipold, D., and Anner, B.M., Endogenous natriuretic factors 7: biospecificity of a natriuretic gamma-tocopherol metabolite LLU-alpha, *J. Pharmacol. Exp. Ther.* 282 (2), 657–662, 1997.

65. Jiang, Q., Elson-Schwab, I., Courtemanche, C., and Ames, B.N., gamma-Tocopherol and its major metabolite, in contrast to alpha-tocopherol, inhibit cyclooxygenase activity in macrophages and epithelial cells, *Proc. Natl. Acad. Sci. U.S.A.* 97 (21), 11494–11499, 2000.

66. Jiang, Q., Christen, S., Shigenaga, M.K., and Ames, B.N., gamma-Tocopherol, the major form of vitamin E in the US diet, deserves more attention, *Am. J. Clin. Nutr.* 74 (6), 714–722, 2001.

67. Jiang, Q. and Ames, B. N., gamma-Tocopherol, but not alpha-tocopherol, decreases proinflammatory eicosanoids and inflammation damage in rats, *FASEB. J.* 17 (8), 816–822, 2003.

68. Galli, F., Stabile, A.M., Betti, M., Conte, C., Pistilli, A., Rende, M., Floridi, A., and Azzi, A., The effect of alpha- and gamma-tocopherol and their carboxyethyl hydroxychroman metabolites on prostate cancer cell proliferation, *Arch Biochem Biophys* 423 (1), 97–102, 2004.

69. Thompson, T.A. and Wilding, G., Androgen antagonist activity by the antioxidant moiety of vitamin E, 2,2,5,7,8-pentamethyl-6-chromanol in human prostate carcinoma cells, *Mol. Cancer. Ther.* 2 (8), 797–803, 2003.

70. Hayashi, T., Kanetoshi, A., Nakamura, M., Tamura, M., and Shirahama, H., Reduction of alpha-tocopherolquinone to alpha-tocopherolhydroquinone in rat hepatocytes, *Biochem. Pharmacol.* 44 (3), 489–493, 1992.

71. Liebler, D.C., Kaysen, K.L., and Kennedy, T.A., Redox cycles of vitamin E: hydrolysis and ascorbic acid dependent reduction of 8a-(alkyldioxy)tocopherones, *Biochemistry* 28 (25), 9772–9777, 1989.

72. Bello, R.I., Kagan, V.E., Tyurin, V., Navarro, F., Alcain, F.J., and Villalba, J.M., Regeneration of lipophilic antioxidants by NAD(P)H:quinone oxidoreductase 1, *Protoplasma* 221 (1–2), 129–135, 2003.

73. Mukai, K., Itoh, S., and Morimoto, H., Stopped-flow kinetic study of vitamin E regeneration reaction with biological hydroquinones (reduced forms of ubiquinone, vitamin K, and tocopherolquinone) in solution, *J Biol Chem* 267 (31), 22277–22281, 1992.

74. Thornton, D.E., Jones, K.H., Jiang, Z., Zhang, H., Liu, G., and Cornwell, D.G., Antioxidant and cytotoxic tocopheryl quinones in normal and cancer cells, *Free Radic. Biol. Med.* 18 (6), 963–976, 1995.

75. Cornwell, D.G., Williams, M.V., Wani, A.A., Wani, G., Shen, E., and Jones, K.H., Mutagenicity of tocopheryl quinones: evolutionary advantage of selective accumulation of dietary alpha-tocopherol, *Nutr. Cancer* 43 (1), 111–118, 2002.

76. Calviello, G., Di Nicuolo, F., Piccioni, E., Marcocci, M.E., Serini, S., Maggiano, N., Jones, K.H., Cornwell, D.G., and Palozza, P., gamma-Tocopheryl quinone induces apoptosis in cancer cells via caspase-9 activation and cytochrome c release, *Carcinogenesis* 24 (3), 427–433, 2003.

77. Soo, C.C., Haqqani, A.S., Hidiroglou, N., Swanson, J.E., Parker, R.S., and Birnboim, H.C., Dose-dependent effects of dietary alpha- and gamma-tocopherols on genetic instability in mouse Mutatect tumors, *J. Natl. Cancer Inst.* 96 (10), 796–800, 2004.

78. Kempnà, P., Cipollone, R., Villacorta, L., Ricciarelli, R., and Zingg, J.M., Isoelectric point mobility shift assay for rapid screening of charged and uncharged ligands bound to proteins, *IUBMB Life* 55, 103–107, 2003.

79. Stocker, A. and Baumann, U., Supernatant protein factor in complex with RRR-alpha-tocopherylquinone: a link between oxidized Vitamin E and cholesterol biosynthesis, *J. Mol. Biol.* 332 (4), 759–765, 2003.

80. Panagabko, C., Morley, S., Hernandez, M., Cassolato, P., Gordon, H., Parsons, R., Manor, D., and Atkinson, J., Ligand specificity in the CRAL-TRIO protein family, *Biochemistry* 42 (21), 6467–6474, 2003.

81. Bauernfeind, J.B., *Tocopherols in Food. Vitamin E. A Comprehensive Treatise*, Marcel Dekker, I., New York and Basel, 1980, 99–167.

82. Bauernfeind, J.C., Rubin, S.H., Surmatis, J.D., and Ofner, A., Carotenoids and fat-soluble vitamins: contribution to food, feed and pharmaceuticals, *Int. Z. Vitaminforsch.* 40 (3), 391–416, 1970.

83. Traber, M.G. and Kayden, H.J., Vitamin E is delivered to cells via the high affinity receptor for low-density lipoprotein, *Am. J. Clin. Nutr.* 40 (4), 747–751, 1984.

84. Handelman, G.J., Epstein, W.L., Peerson, J., Spiegelman, D., Machlin, L.J., and Dratz, E.A., Human adipose alpha-tocopherol and gamma-tocopherol kinetics during and after 1 y of alpha-tocopherol supplementation, *Am. J. Clin. Nutr.* 59 (5), 1025–1032, 1994.

85. Hosomi, A., Arita, M., Sato, Y., Kiyose, C., Ueda, T., Igarashi, O., Arai, H., and Inoue, K., Affinity for alpha-tocopherol transfer protein as a determinant of the biological activities of vitamin E analogs, *FEBS. Lett.* 409 (1), 105–108, 1997.

86. Mardones, P. and Rigotti, A., Cellular mechanisms of vitamin E uptake: relevance in alpha-tocopherol metabolism and potential implications for disease, *J. Nutr. Biochem.* 15 (5), 252–260, 2004.

87. Kolleck, I., Schlame, M., Fechner, H., Looman, A.C., Wissel, H., and Rustow, B., HDL is the major source of vitamin E for type II pneumocytes, *Free Radic. Biol. Med.* 27 (7–8), 882–890., 1999.

88. Goti, D., Hrzenjak, A., Levak-Frank, S., Frank, S., van Der Westhuyzen, D.R., Malle, E., and Sattler, W., Scavenger receptor class B, type I is expressed in porcine brain capillary endothelial cells and contributes to selective uptake of HDL- associated vitamin E, *J. Neurochem.* 76 (2), 498–508., 2001.

89. Goti, D., Reicher, H., Malle, E., Kostner, G.M., Panzenboeck, U., and Sattler, W., High-density lipoprotein (HDL3)-associated alpha-tocopherol is taken up by HepG2 cells via the selective uptake pathway and resecreted with endogenously synthesized apo-lipoprotein B-rich lipoprotein particles, *Biochem. J.* 332 (Pt 1), 57–65, 1998.

90. Copp, R.P., Wisniewski, T., Hentati, F., Larnaout, A., Ben Hamida, M., and Kayden, H.J., Localization of alpha-tocopherol transfer protein in the brains of patients with ataxia with vitamin E deficiency and other oxidative stress related neurodegenerative disorders, *Brain. Res.* 822 (1–2), 80–87, 1999.

91. Yokota, T., Shiojiri, T., Gotoda, T., Arita, M., Arai, H., Ohga, T., Kanda, T., Suzuki, J., Imai, T., Matsumoto, H., Harino, S., Kiyosawa, M., Mizusawa, H., and Inoue, K., Friedreich-like ataxia with retinitis pigmentosa caused by the His101Gln mutation of the alpha-tocopherol transfer protein gene, *Ann. Neurol.* 41 (6), 826–832, 1997.

92. Tamaru, Y., Hirano, M., Kusaka, H., Ito, H., Imai, T., and Ueno, S., alpha-Tocopherol transfer protein gene: exon skipping of all transcripts causes ataxia, *Neurology* 49 (2), 584–588, 1997.

93. Jishage, K., Arita, M., Igarashi, K., Iwata, T., Watanabe, M., Ogawa, M., Ueda, O., Kamada, N., Inoue, K., Arai, H., and Suzuki, H., Alpha-tocopherol transfer protein is important for the normal development of placental labyrinthine trophoblasts in mice, *J. Biol. Chem.* 276 (3), 1669–1672., 2001.

94. Federico, A., Ataxia with isolated vitamin E deficiency: a treatable neurologic disorder resembling Friedreich's ataxia, *Neurol. Sci.* 25 (3), 119–121, 2004.

95. Traber, M.G., Burton, G.W., Ingold, K.U., and Kayden, H.J., RRR- and SRR-alpha-tocopherols are secreted without discrimination in human chylomicrons, but RRR-alpha-tocopherol is preferentially secreted in very low density lipoproteins, *J. Lipid. Res.* 31 (4), 675–685, 1990.

96. Traber, M.G., Sokol, R.J., Kohlschutter, A., Yokota, T., Muller, D.P., Dufour, R., and Kayden, H.J., Impaired discrimination between stereoisomers of alpha-tocopherol in patients with familial isolated vitamin E deficiency, *J. Lipid. Res.* 34 (2), 201–210, 1993.

97. Hodis, H.N., Mack, W.J., LaBree, L., Mahrer, P.R., Sevanian, A., Liu, C.R., Liu, C.H., Hwang, J., Selzer, R.H., and Azen, S.P., Alpha-tocopherol supplementation in healthy individuals reduces low-density lipoprotein oxidation but not atherosclerosis: the Vitamin E Atherosclerosis Prevention Study (VEAPS), *Circulation* 106 (12), 1453–1459, 2002.

98. Leonard, S.W., Terasawa, Y., Farese, R.V., Jr., and Traber, M.G., Incorporation of deuterated RRR- or all-rac-alpha-tocopherol in plasma and tissues of alpha-toco-pherol transfer protein — null mice, *Am. J. Clin. Nutr.* 75 (3), 555–560, 2002.

99. Ham, A.J. and Liebler, D.C., Vitamin E oxidation in rat liver mitochondria, *Biochemistry* 34 (17), 5754–5761, 1995.

100. Tran, K. and Chan, A.C., Comparative uptake of alpha- and gamma-tocopherol by human endothelial cells, *Lipids* 27 (1), 38–41, 1992.

101. Catignani, G.L., An alpha-tocopherol binding protein in rat liver cytoplasm, *Biochem. Biophys. Res. Commun.* 67 (1), 66–72., 1975.

102. Dutta-Roy, A.K., Leishman, D.J., Gordon, M.J., Campbell, F.M., and Duthie, G.G., Identification of a low molecular mass (14.2 kDa) alpha-tocopherol-binding protein in the cytosol of rat liver and heart, *Biochem. Biophys. Res. Commun.* 196 (3), 1108–1112, 1993.

103. Dutta-Roy, A.K., Alpha-tocopherol-binding proteins: purification and characterization, *Methods Enzymol.* 282, 278–297, 1997.

104. Dutta-Roy, A.K., Molecular mechanism of cellular uptake and intracellular translocation of alpha-tocopherol: role of tocopherol-binding proteins, *Food. Chem. Toxicol.* 37 (9–10), 967–971, 1999.

105. Kaempf-Rotzoll, D.E., Traber, M.G., and Arai, H., Vitamin E and transfer proteins, *Curr. Opin. Lipidol.* 14 (3), 249–254, 2003.

106. Zimmer, S., Stocker, A., Sarbolouki, M.N., Spycher, S.E., Sassoon, J., and Azzi, A., A novel human tocopherol-associated protein: cloning, in vitro expression, and characterization, *J. Biol. Chem.* 275 (33), 25672–25680, 2000.

107. Shibata, N., Arita, M., Misaki, Y., Dohmae, N., Takio, K., Ono, T., Inoue, K., and Arai, H., Supernatant protein factor, which stimulates the conversion of squalene to lanosterol, is a cytosolic squalene transfer protein and enhances cholesterol biosynthesis, *Proc. Natl. Acad. Sci. U.S.A.* 98 (5), 2244–2249., 2001.

108. Bradford, A., Atkinson, J., Fuller, N., and Rand, R.P., The effect of vitamin E on the structure of membrane lipid assemblies, *J. Lipid. Res.*, 44, 1940–1945, 2003.

109. Kempnà, P., Zingg, J.M., Ricciarelli, R., Hierl, M., Saxena, S., and Azzi, A., Cloning of novel human SEC14p-like proteins: cellular localization, ligand binding and functional properties, *Free Radic. Biol. Med.* 34, 1458–1472, 2003.

110. Anantharaman, V. and Aravind, L., The GOLD domain, a novel protein module involved in Golgi function and secretion, *Genome Biol.* 3 (5), research 0023, 2002.

111. Caras, I.W., Friedlander, E.J., and Bloch, K., Interactions of supernatant protein factor with components of the microsomal squalene epoxidase system. Binding of supernatant protein factor to anionic phospholipids, *J. Biol. Chem.* 255 (8), 3575–3580, 1980.

112. Azzi, A., Ricciarelli, R., and Zingg, J.M., Non-antioxidant molecular functions of alpha-tocopherol (vitamin E), *FEBS Lett.* 519 (1–3), 8–10, 2002.

113. Hacquebard, M. and Carpentier, Y.A., Vitamin E: absorption, plasma transport and cell uptake, *Curr. Opin. Clin. Nutr. Metab. Care* 8 (2), 133–138, 2005.

114. Oram, J.F., Vaughan, A.M., and Stocker, R., ATP-binding cassette transporter A1 mediates cellular secretion of alpha-tocopherol, *J. Biol. Chem.* 276 (43), 39898–39902, 2001.

115. Halliwell, B., Rafter, J., and Jenner, A., Health promotion by flavonoids, tocopherols, tocotrienols, and other phenols: direct or indirect effects? Antioxidant or not? *Am. J. Clin. Nutr.* 81 (1 Suppl), 268S–276S, 2005.

116. Wang, X. and Quinn, P.J., The location and function of vitamin E in membranes (review), *Mol. Membr. Biol.* 17 (3), 143–156, 2000.

117. Yoshida, Y., Niki, E., and Noguchi, N., Comparative study on the action of tocopherols and tocotrienols as antioxidant: chemical and physical effects, *Chem. Phys. Lipids* 123 (1), 63–75, 2003.

118. Suzuki, Y.J., Tsuchiya, M., Wassall, S.R., Choo, Y.M., Govil, G., Kagan, V.E., and Packer, L., Structural and dynamic membrane properties of alpha-tocopherol and alpha-tocotrienol: implication to the molecular mechanism of their antioxidant potency, *Biochemistry* 32 (40), 10692–10699, 1993.

119. Pearce, B.C., Parker, R.A., Deason, M.E., Qureshi, A.A., and Wright, J.J., Hypocholesterolemic activity of synthetic and natural tocotrienols, *J. Med. Chem.* 35 (20), 3595–3606, 1992.

120. Parker, R.A., Pearce, B.C., Clark, R.W., Gordon, D.A., and Wright, J.J., Tocotrienols regulate cholesterol production in mammalian cells by post-transcriptional suppression of 3-hydroxy-3-methylglutaryl-coenzyme A reductase, *J. Biol. Chem.* 268 (15), 11230–11238, 1993.

121. Wang, Q., Theriault, A., Gapor, A., and Adeli, K., Effects of tocotrienol on the intracellular translocation and degradation of apolipoprotein B: possible involvement of a proteasome independent pathway, *Biochem. Biophys. Res. Commun.* 246 (3), 640–643, 1998.

122. Khor, H.T. and Ng, T.T., Effects of administration of alpha-tocopherol and tocotrienols on serum lipids and liver HMG CoA reductase activity, *Int. J. Food Sci. Nutr.* 51 Suppl, S3–11, 2000.

123. Black, T.M., Wang, P., Maeda, N., and Coleman, R.A., Palm tocotrienols protect ApoE +/- mice from diet-induced atheroma formation, *J. Nutr.* 130 (10), 2420–2426, 2000.

124. Khanna, S., Roy, S., Ryu, H., Bahadduri, P., Swaan, P.W., Ratan, R.R., and Sen, C.K., Molecular basis of vitamin E action. Tocotrienol modulates 12-lipoxygenase, a key mediator of glutamate-induced neurodegeneration, *J. Biol. Chem.* 2003.

125. Das, S., Powell, S.R., Wang, P., Divald, A., Nesaretnam, K., Tosaki, A., Cordis, G.A., Maulik, N., and Das, D.K., Cardioprotection with palm tocotrienol: antioxidant activity of tocotrienol is linked with its ability to stabilize proteasomes, *Am. J. Physiol. Heart Circ. Physiol.*, 289(1), H361–367, 2005.

126. Osakada, F., Hashino, A., Kume, T., Katsuki, H., Kaneko, S., and Akaike, A., Alpha-tocotrienol provides the most potent neuroprotection among vitamin E analogs on cultured striatal neurons, *Neuropharmacology* 47 (6), 904–915, 2004.

127. Theriault, A., Chao, J.T., Wang, Q., Gapor, A., and Adeli, K., Tocotrienol: a review of its therapeutic potential, *Clin. Biochem.* 32 (5), 309–319, 1999.

128. Theriault, A., Chao, J.T., and Gapor, A., Tocotrienol is the most effective vitamin E for reducing endothelial expression of adhesion molecules and adhesion to monocytes, *Atherosclerosis* 160 (1), 21–30, 2002.

129. Noguchi, N., Hanyu, R., Nonaka, A., Okimoto, Y., and Kodama, T., Inhibition of THP-1 cell adhesion to endothelial cells by alpha-tocopherol and alpha-tocotrienol is dependent on intracellular concentration of the antioxidants, *Free Radic. Biol. Med.* 34 (12), 1614–1620, 2003.

130. Yu, W., Simmons-Menchaca, M., Gapor, A., Sanders, B.G., and Kline, K., Induction of apoptosis in human breast cancer cells by tocopherols and tocotrienols, *Nutr. Cancer* 33 (1), 26–32, 1999.

131. McIntyre, B.S., Briski, K.P., Tirmenstein, M.A., Fariss, M.W., Gapor, A., and Sylvester, P.W., Antiproliferative and apoptotic effects of tocopherols and tocotrienols on normal mouse mammary epithelial cells, *Lipids* 35 (2), 171–180, 2000.

132. McIntyre, B.S., Briski, K.P., Gapor, A., and Sylvester, P.W., Antiproliferative and apoptotic effects of tocopherols and tocotrienols on preneoplastic and neoplastic mouse mammary epithelial cells, *Proc. Soc. Exp. Biol. Med.* 224 (4), 292–301, 2000.

133. Sylvester, P.W., McIntyre, B.S., Gapor, A., and Briski, K.P., Vitamin E inhibition of normal mammary epithelial cell growth is associated with a reduction in protein kinase C(alpha) activation, *Cell. Prolif.* 34 (6), 347–357, 2001.

134. Komiyama, K., Iizuka, K., Yamaoka, M., Watanabe, H., Tsuchiya, N., and Umezawa, I., Studies on the biological activity of tocotrienols, *Chem. Pharm. Bull. (Tokyo)* 37 (5), 1369–1371, 1989.

135. Sylvester, P.W. and Shah, S.J., Mechanisms mediating the antiproliferative and apoptotic effects of vitamin E in mammary cancer cells, *Front Biosci.* 10, 699–709, 2005.

136. Sies, H., Oxidative stress: oxidants and antioxidants, *Exp. Physiol.* 82 (2), 291–295, 1997.

137. Meydani, M., Vitamin E and atherosclerosis: beyond prevention of LDL oxidation, *J. Nutr.* 131 (2), 366S–368S, 2001.

138. Jialal, I., Devaraj, S., and Kaul, N., The effect of alpha-tocopherol on monocyte proatherogenic activity, *J. Nutr.* 131 (2), 389S–394S, 2001.

139. Packer, L., Weber, S.U., and Rimbach, G., Molecular aspects of alpha-tocotrienol antioxidant action and cell signalling, *J. Nutr.* 131 (2), 369S–373S., 2001.

140. Mashima, R., Witting, P.K., and Stocker, R., Oxidants and antioxidants in atherosclerosis, *Curr. Opin. Lipidol.* 12 (4), 411–418, 2001.

141. Asplund, K., Antioxidant vitamins in the prevention of cardiovascular disease: a systematic review, *J. Intern. Med.* 251 (5), 372–392, 2002.

142. Upston, J.M., Kritharides, L., and Stocker, R., The role of vitamin E in atherosclerosis, *Prog. Lipid. Res.* 42 (5), 405–422, 2003.

143. Traber, M.G. and Packer, L., Vitamin E: beyond antioxidant function, *Am. J. Clin. Nutr.* 62 (6 Suppl), 1501S–1509S, 1995.

144. Boscoboinik, D., Szewczyk, A., Hensey, C., and Azzi, A., Inhibition of cell proliferation by alpha-tocopherol. Role of protein kinase C, *J. Biol. Chem.* 266 (10), 6188–6194, 1991.

145. Ricciarelli, R., Tasinato, A., Clement, S., Ozer, N.K., Boscoboinik, D., and Azzi, A., alpha-Tocopherol specifically inactivates cellular protein kinase C alpha by changing its phosphorylation state, *Biochem. J.* 334 (Pt1), 243–249, 1998.

146. Ricciarelli, R. and Azzi, A., Regulation of recombinant PKC alpha activity by protein phosphatase 1 and protein phosphatase 2A, *Arch. Biochem. Biophys.* 355 (2), 197–200, 1998.

147. Shah, S.J. and Sylvester, P.W., Gamma-tocotrienol inhibits neoplastic mammary epithelial cell proliferation by decreasing Akt and nuclear factor kappaB activity, *Exp. Biol. Med. (Maywood)* 230 (4), 235–241, 2005.

148. Pentland, A.P., Morrison, A.R., Jacobs, S.C., Hruza, L.L., Hebert, J.S., and Packer, L., Tocopherol analogs suppress arachidonic acid metabolism via phospholipase inhibition, *J. Biol. Chem.* 267 (22), 15578–15584, 1992.

149. Abate, A., Yang, G., Dennery, P.A., Oberle, S., and Schroder, H., Synergistic inhibition of cyclooxygenase-2 expression by vitamin E and aspirin, *Free Radic. Biol. Med.* 29 (11), 1135–1142, 2000.

150. Chandra, V., Jasti, J., Kaur, P., Betzel, C., Srinivasan, A., and Singh, T.P., First structural evidence of a specific inhibition of phospholipase A2 by alpha-tocopherol (vitamin E) and its implications in inflammation: crystal structure of the complex formed between phospholipase A2 and alpha-tocopherol at 1.8 A resolution, *J. Mol. Biol.* 320 (2), 215–222, 2002.

151. Min, J., Guo, J., Zhao, F., and Cai, D., Effect of apoptosis induced by different vitamin E homologous analogues in human hepatoma cells (HepG2), *Wei Sheng Yan Jiu* 32 (2), 131–133, 2003.

152. Azzi, A., Gysin, R., Kempna, P., Ricciarelli, R., Villacorta, L., Visarius, T., and Zingg, J.M., Regulation of gene and protein expression by vitamin E, *Free Radic. Res.* 36 (1), 30–35, 2002.

153. Landes, N., Pfluger, P., Kluth, D., Birringer, M., Ruhl, R., Bol, G.F., Glatt, H., and Brigelius-Flohe, R., Vitamin E activates gene expression via the pregnane X receptor, *Biochem. Pharmacol.* 65 (2), 269–273, 2003.

154. Yamauchi, J., Iwamoto, T., Kida, S., Masushige, S., Yamada, K., and Esashi, T., Tocopherol-associated protein is a ligand-dependent transcriptional activator, *Biochem. Biophys. Res. Commun.* 285 (2), 295–299., 2001.

155. Campbell, S.E., Stone, W.L., Whaley, S.G., Qui, M., and Krishnan, K., Gamma (gamma) tocopherol upregulates peroxisome proliferator activated receptor (PPAR) gamma (gamma) expression in SW 480 human colon cancer cell lines, *BMC Cancer* 3 (1), 25, 2003.

156. Traber, M.G., Vitamin E, nuclear receptors and xenobiotic metabolism, *Arch. Biochem. Biophys.* 423 (1), 6–11, 2004.

157. Gimeno, A., Zaragoza, R., Vina, J.R., and Miralles, V.J., Vitamin E activates CRABP-II gene expression in cultured human fibroblasts, role of protein kinase C, *FEBS Lett* 569 (1–3), 240–244, 2004.

158. Stauble, B., Boscoboinik, D., Tasinato, A., and Azzi, A., Modulation of activator protein-1 (AP-1) transcription factor and protein kinase C by hydrogen peroxide and D-alpha-tocopherol in vascular smooth muscle cells, *Eur. J. Biochem.* 226 (2), 393–402, 1994.

159. Azzi, A., Boscoboinik, D., Marilley, D., Ozer, N.K., Stauble, B., and Tasinato, A., Vitamin E: a sensor and an information transducer of the cell oxidation state, *Am. J. Clin. Nutr.* 62 (6 Suppl), 1337S–1346S, 1995.

160. Yokota, T., Uchihara, T., Kumagai, J., Shiojiri, T., Pang, J.J., Arita, M., Arai, H., Hayashi, M., Kiyosawa, M., Okeda, R., and Mizusawa, H., Postmortem study of ataxia with retinitis pigmentosa by mutation of the alpha-tocopherol transfer protein gene, *J. Neurol. Neurosurg. Psychiatry* 68 (4), 521–525, 2000.

161. Jauslin, M.L., Meier, T., Smith, R.A., and Murphy, M.P., Mitochondria-targeted antioxidants protect Friedreich ataxia fibroblasts from endogenous oxidative stress more effectively than untargeted antioxidants, *Faseb J* 17 (13), 1972–1974, 2003.

162. Hart, P.E., Lodi, R., Rajagopalan, B., Bradley, J.L., Crilley, J. G., Turner, C., Blamire, A.M., Manners, D., Styles, P., Schapira, A.H., and Cooper, J.M., Antioxidant treatment of patients with Friedreich ataxia: four-year follow-up, *Arch Neurol* 62 (4), 621–626, 2005.

163. Meydani, M., Koga, T., and Ali, S., Vitamin E deficiency, in *Encyclopedia of Life Sciences,* Nature Publishing Group, 2001, 1–6.

164. Salonen, J.T., Nyyssonen, K., Salonen, R., Lakka, H.M., Kaikkonen, J., Porkkala-Sarataho, E., Voutilainen, S., Lakka, T.A., Rissanen, T., Leskinen, L., Tuomainen, T.P., Valkonen, V.P., Ristonmaa, U., and Poulsen, H.E., Antioxidant Supplementation in Atherosclerosis Prevention (ASAP) study: a randomized trial of the effect of vitamins E and C on 3-year progression of carotid atherosclerosis, *J. Intern. Med.* 248 (5), 377–386., 2000.

165. Kaul, N., Devaraj, S., and Jialal, I., Alpha-tocopherol and atherosclerosis, *Exp. Biol. Med. (Maywood)* 226 (1), 5–12, 2001.

166. Azzi, A., Gysin, R., Kempna, P., Ricciarelli, R., Villacorta, L., Visarius, T., and Zingg, J.M., The role of alpha-tocopherol in preventing disease: from epidemiology to molecular events, *Mol. Aspects Med.* 24 (6), 325–336, 2003.

167. Munteanu, A., Zingg, J.M., and Azzi, A., Anti-atherosclerotic effects of vitamin E — myth or reality? *J. Cell Mol. Med.* 8 (1), 59–76, 2004.

168. Hathcock, J.N., Azzi, A., Blumberg, J., Bray, T., Dickinson, A., Frei, B., Jialal, I., Johnston, C.S., Kelly, F.J., Kraemer, K., Packer, L., Parthasarathy, S., Sies, H., and Traber, M.G., Vitamins E and C are safe across a broad range of intakes, *Am. J. Clin. Nutr.* 81 (4), 736–745, 2005.

169. Brown, B.G. and Crowley, J., Is there any hope for vitamin E?, *Jama* 293 (11), 1387–1390, 2005.

170. Lonn, E., Bosch, J., Yusuf, S., Sheridan, P., Pogue, J., Arnold, J.M., Ross, C., Arnold, A., Sleight, P., Probstfield, J., and Dagenais, G.R., Effects of long-term vitamin E supplementation on cardiovascular events and cancer: a randomized controlled trial, *JAMA* 293 (11), 1338–1347, 2005.

171. Albanes, D., Heinonen, O.P., Huttunen, J.K., Taylor, P.R., Virtamo, J., Edwards, B.K., Haapakoski, J., Rautalahti, M., Hartman, A.M., Palmgren, J. et al., Effects of alpha-tocopherol and beta-carotene supplements on cancer incidence in the Alpha-Tocopherol Beta-Carotene Cancer Prevention Study, *Am. J. Clin. Nutr.* 62 (6 Suppl), 1427S–1430S, 1995.

172. Woodson, K., Tangrea, J.A., Barrett, M.J., Virtamo, J., Taylor, P.R., and Albanes, D., Serum alpha-tocopherol and subsequent risk of lung cancer among male smokers, *J. Natl. Cancer. Inst.* 91 (20), 1738–1743, 1999.

173. Leppala, J.M., Virtamo, J., Fogelholm, R., Huttunen, J.K., Albanes, D., Taylor, P.R., and Heinonen, O.P., Controlled trial of alpha-tocopherol and beta-carotene supplements on stroke incidence and mortality in male smokers, *Arterioscler. Thromb. Vasc. Biol.* 20 (1), 230–235, 2000.

174. Albanes, D., Malila, N., Taylor, P.R., Huttunen, J.K., Virtamo, J., Edwards, B.K., Rautalahti, M., Hartman, A.M., Barrett, M.J., Pietinen, P., Hartman, T.J., Sipponen, P., Lewin, K., Teerenhovi, L., Hietanen, P., Tangrea, J.A., Virtanen, M., and Heinonen, O.P., Effects of supplemental alpha-tocopherol and beta-carotene on colorectal cancer: results from a controlled trial (Finland), *Cancer Causes Control* 11 (3), 197–205, 2000.

175. Giovannucci, E., Gamma-tocopherol: a new player in prostate cancer prevention? *J. Natl. Cancer Inst.* 92 (24), 1966–1967, 2000.

176. Houglum, K., Venkataramani, A., Lyche, K., and Chojkier, M., A pilot study of the effects of d-alpha-tocopherol on hepatic stellate cell activation in chronic hepatitis C, *Gastroenterology* 113 (4), 1069–1073, 1997.

177. Harrison, S.A., Torgerson, S., Hayashi, P., Ward, J., and Schenker, S., Vitamin E and vitamin C treatment improves fibrosis in patients with nonalcoholic steatohepatitis, *Am J Gastroenterol* 98 (11), 2485–2490, 2003.

178. Berman, K. and Brodaty, H., Tocopherol (vitamin E) in Alzheimer's disease and other neurodegenerative disorders, *CNS Drugs* 18 (12), 807–825, 2004.

179. Morris, M.C., Evans, D.A., Tangney, C.C., Bienias, J.L., Wilson, R.S., Aggarwal, N.T., and Scherr, P. A., Relation of the tocopherol forms to incident Alzheimer disease and to cognitive change, *Am. J. Clin. Nutr.* 81 (2), 508–514, 2005.

180. Fariss, M.W. and Zhang, J.G., Vitamin E therapy in Parkinson's disease, *Toxicology* 189 (1–2), 129–46, 2003.

181. Doring, F., Rimbach, G., and Lodge, J.K., In silico search for single nucleotide polymorphisms in genes important in vitamin E homeostasis, *IUBMB Life* 56 (10), 615–620, 2004.

182. Barbeau, A., Friedreich's ataxia 1980. An overview of the physiopathology, *Can. J. Neurol. Sci.* 7 (4), 455–468, 1980.

183. Hirai, K., Aliev, G., Nunomura, A., Fujioka, H., Russell, R.L., Atwood, C.S., Johnson, A.B., Kress, Y., Vinters, H.V., Tabaton, M., Shimohama, S., Cash, A.D., Siedlak, S.L., Harris, P.L., Jones, P.K., Petersen, R.B., Perry, G., and Smith, M.A., Mitochondrial abnormalities in Alzheimer's disease, *J. Neurosci.* 21 (9), 3017–3023, 2001.
184. Kaplan, J., Spinocerebellar ataxias due to mitochondrial defects, *Neurochem. Int.* 40 (6), 553–557, 2002.
185. Marriage, B., Cladinin, M.T., and Glerum, D.M., Nutritional cofactor treatment in mitochondrial disorders, *J. Am. Diet. Assoc.* 103 (8), 1029–1038, 2003.
186. Dial, S. and Eitenmiller, R.R., Tocopherols and tocotrienols in key foods in the U.S. diet, in *Nutrition, Lipids, Health, and Disease*, Ong, A.S.F., Niki, E., and Packer, L., Eds., AOCS Press, 1995, 327–342.
187. Fuchs, J., Weber, S., Podda, M., Groth, N., Herrling, T., Packer, L., and Kaufmann, R., HPLC analysis of vitamin E isoforms in human epidermis: correlation with minimal erythema dose and free radical scavenging activity, *Free Radic. Biol. Med.* 34 (3), 330–336, 2003.
188. Bjorneboe, A., Bjorneboe, G.E., and Drevon, C.A., Serum half-life, distribution, hepatic uptake and biliary excretion of alpha-tocopherol in rats, *Biochim. Biophys. Acta* 921 (2), 175–181, 1987.
189. Mustacich, D.J., Shields, J., Horton, R.A., Brown, M.K., and Reed, D.J., Biliary secretion of alpha-tocopherol and the role of the mdr2 P-glycoprotein in rats and mice, *Arch. Biochem. Biophys.* 350 (2), 183–192, 1998.
190. Traber, M.G. and Arai, H., Molecular mechanisms of vitamin E transport, *Annu. Rev. Nutr.* 19, 343–355, 1999.
191. Caras, I.W. and Bloch, K., Effects of a supernatant protein activator on microsomal squalene-2,3-oxide-lanosterol cyclase, *J. Biol. Chem.* 254 (23), 11816–11821, 1979.
192. Porter, T.D., Supernatant protein factor and tocopherol-associated protein: an unexpected link between cholesterol synthesis and vitamin E (review), *J. Nutr. Biochem.* 14 (1), 3–6, 2003.
193. Heiser, M., Hutter-Paier, B., Jerkovic, L., Pfragner, R., Windisch, M., Becker-Andre, M., and Dieplinger, H., Vitamin E binding protein afamin protects neuronal cells in vitro, *J. Neural. Transm. Suppl.* (62), 337–345, 2002.
194. Voegele, A.F., Jerkovic, L., Wellenzohn, B., Eller, P., Kronenberg, F., Liedl, K.R., and Dieplinger, H., Characterization of the vitamin E-binding properties of human plasma afamin, *Biochemistry* 41 (49), 14532–14538, 2002.
195. Cavalier, L., Ouahchi, K., Kayden, H.J., Di Donato, S., Reutenauer, L., Mandel, J.L., and Koenig, M., Ataxia with isolated vitamin E deficiency: heterogeneity of mutations and phenotypic variability in a large number of families, *Am. J. Hum. Genet.* 62 (2), 301–310, 1998.
196. Benomar, A., Yahyaoui, M., Marzouki, N., Birouk, N., Bouslam, N., Belaidi, H., Amarti, A., Ouazzani, R., and Chkili, T., Vitamin E deficiency ataxia associated with adenoma, *J. Neurol. Sci.* 162 (1), 97–101., 1999.
197. Bowry, V.W., Ingold, K.U., and Stocker, R., Vitamin E in human low-density lipoprotein. When and how this antioxidant becomes a pro-oxidant, *Biochem. J.* 288 (Pt 2), 341–344, 1992.
198. Upston, J.M., Terentis, A.C., and Stocker, R., Tocopherol-mediated peroxidation of lipoproteins: implications for vitamin E as a potential antiatherogenic supplement, *FASEB. J.* 13 (9), 977–994., 1999.
199. Wolf, G., gamma-Tocopherol: an efficient protector of lipids against nitric oxide-initiated peroxidative damage, *Nutr. Rev.* 55 (10), 376–378, 1997.

200. Christen, S., Woodall, A.A., Shigenaga, M.K., Southwell-Keely, P.T., Duncan, M.W., and Ames, B. N., Gamma-tocopherol traps mutagenic electrophiles such as NO(X) and complements alpha-tocopherol: physiological implications, *Proc. Natl. Acad. Sci. U.S.A.* 94 (7), 3217–3222., 1997.

201. Goss, S.P., Hogg, N., and Kalyanaraman, B., The effect of alpha-tocopherol on the nitration of gamma-tocopherol by peroxynitrite, *Arch. Biochem. Biophys.* 363 (2), 333–340, 1999.

202. Sen, C.K., Khanna, S., Roy, S., and Packer, L., Molecular basis of vitamin E action. Tocotrienol potently inhibits glutamate-induced pp60(c-Src) kinase activation and death of HT4 neuronal cells, *J. Biol. Chem.* 275 (17), 13049–13055, 2000.

203. Venugopal, S.K., Devaraj, S., and Jialal, I., RRR-alpha-tocopherol decreases the expression of the major scavenger receptor, CD36, in human macrophages via inhibition of tyrosine kinase (Tyk2), *Atherosclerosis* 175 (2), 213–220, 2004.

204. Tran, K. and Chan, A.C., R,R,R-alpha-tocopherol potentiates prostacyclin release in human endothelial cells. Evidence for structural specificity of the tocopherol molecule, *Biochim. Biophys. Acta* 1043 (2), 189–197, 1990.

205. Devaraj, S. and Jialal, I., Alpha-tocopherol decreases interleukin-1 beta release from activated human monocytes by inhibition of 5-lipoxygenase, *Arterioscler. Thromb. Vasc. Biol.* 19 (4), 1125–1133., 1999.

206. van Haaften, R.I., Haenen, G.R., van Bladeren, P.J., Bogaards, J.J., Evelo, C.T., and Bast, A., Inhibition of various glutathione S-transferase isoenzymes by RRR-alpha-tocopherol, *Toxicol. in Vitro* 17 (3), 245–251, 2003.

207. Cachia, O., Benna, J.E., Pedruzzi, E., Descomps, B., Gougerot-Pocidalo, M.A., and Leger, C.L., Alpha-tocopherol inhibits the respiratory burst in human monocytes. Attenuation of p47(phox) membrane translocation and phosphorylation, *J. Biol. Chem.* 273 (49), 32801–32805, 1998.

208. Gohil, K., Schock, B.C., Chakraborty, A.A., Terasawa, Y., Raber, J., Farese, R.V., Jr., Packer, L., Cross, C.E., and Traber, M.G., Gene expression profile of oxidant stress and neurodegeneration in transgenic mice deficient in alpha-tocopherol transfer protein, *Free Radic. Biol. Med.* 35 (11), 1343–1354, 2003.

209. Barella, L., Muller, P.Y., Schlachter, M., Hunziker, W., Stocklin, E., Spitzer, V., Meier, N., de Pascual-Teresa, S., Minihane, A. M., and Rimbach, G., Identification of hepatic molecular mechanisms of action of alpha-tocopherol using global gene expression profile analysis in rats, *Biochim. Biophys. Acta* 1689 (1), 66–74, 2004.

Index